VIRUSES, ALLERGIES AND THE IMMUNE SYSTEM

JAN DE VRIES was born in 1937 in Holland and grew up during the difficult war years in occupied territory. Although he graduated in pharmacy, he soon turned to alternative medicine. His most influential teacher was Dr Alfred Vogel in Switzerland, and they have worked together closely for 35 years.

In 1970 he and his family moved to Scotland and settled on the west coast in Troon, where he set up a residential clinic called Mokoia. He also has clinics in Newcastle, Edinburgh and London. Since 1990 he has been involved in Klein Vink in Arcen, Holland, doing research into the efficacy of herbal medicine for the European Commission.

He lectures throughout the world and is a regular broadcaster on BBC radio.

Books available from the same author

By Appointment Only series
Arthritis, Rheumatism and Psoriasis
Asthma and Bronchitis
Cancer and Leukaemia
Heart and Blood Circulatory Problems
Migraine and Epilepsy
The Miracle of Life
Multiple Sclerosis
Neck and Back Problems
Realistic Weight Control
Skin Diseases
Stomach and Bowel Disorders
Stress and Nervous Disorders
Traditional Home and Herbal Remedies
Viruses, Allergies and the Immune System

Nature's Gift series
Air – The Breath of Life
Body Energy
Food
Water – Healer or Poison?

Well Woman series
Menopause
Menstrual and Pre-Menstrual Tension
Pregnancy and Childbirth

The Jan de Vries Healthcare series
How to Live a Healthy Life – A Handbook to Better Health
Questions and Answers on Family Health
The Five Senses

Also available from the same author
Life Without Arthritis – The Maori Way
Who's Next?

Viruses, Allergies and the Immune System

JAN DE VRIES

First published in Great Britain in 1988 by
MAINSTREAM PUBLISHING COMPANY (EDINBURGH) LTD
7 Albany Street
Edinburgh EH1 3UG

Reprinted 1989, 1990, 1995, 1997

ISBN 1 85158 176 6

A catalogue record for this book is available from the British Library

Reproduced from disc in 10.5/12 Palatino by Polyprint, 48 Pleasance, Edinburgh, EH8 9TJ
Printed and bound in Finland by WSOY

Contents

Foreword

IMMUNITY IS THE body's capacity to distinguish self from non-self, or foreign matter, and to eliminate or neutralise the latter. This definition by the United States National Institute of Health is further expanded by an explanation of the two interrelated aspects of the means through which this goal is reached, namely, the immune system. First is the formation of antibodies, protein molecules produced in response to specific antigens or foreign bodies, which have the capacity to neutralise those harmful invaders whether they be bacteria or viruses. Second is the sensitisation of a variety of pre-existing cells by the harmful antigens. Such cells are B-lymphocytes, which are precursors of the cells that produce antibodies; T-lymphocytes, which induce antibody formation and are needed for cellular immunity; and monocytes, specific white blood cells which become scavengers of undesirable cells and foreign bodies that enter the system.

To the more obvious unwanted infiltrators, like harmful bacteria and viruses, we must now add undesirable pollutants, irradiation, allergy-inducing substances and food additives that in many cases have not been proved to be harmless to everyone's satisfaction. Of great concern, as our knowledge grows of them, is the formation within the body cells themselves of harmful entities like free radicals and cancer-producing compounds that have to be destroyed or rendered harmless by existing body mechanisms. The human organism must also have to cope with foreign substances deliberately introduced into it, for example, medicinal drugs, tobacco smoke and alcohol, or unwittingly, as in food contaminants like pesticide residues, traces of fertilisers and toxic heavy metals from the soil and other sources. The immune system has all of these undesirable factors to cope with and usually does so admirably. In the past, an efficient immune system has been regarded as a result of chance, heredity and, perhaps, lifestyle. Now it is becoming apparent that the effectiveness of the vastly complex protective mechanisms of the body depend very much upon the factors that control those mechanisms and these factors are in the main nutritional in origin. A prime example of this is in the relationship of micro-nutrients to cancer prevention and a brief review of this illustrates an exciting development of great potential benefit.

In the United States, cancer is responsible for more than 20 per cent of the approximate annual death rate of two million. Similar proportions apply also to the West in general. Thirty per cent of all cancer deaths is attributable to cigarette smoking. Heavy alcohol intakes account for about three per cent of all cancer deaths. Hence one-third of deaths from cancer can be avoided by desisting from these social (or perhaps anti-social) practices. A further third of all cancer deaths is believed to result from dietary factors, but which ones are responsible is still unclear although the known production of specific carcinogenic substances in and from the diet must play a part. These are facts, but on the bright side is the emerging evidence that diet may also be important in preventing the onset of cancer.

Modern epidemiological evidence and intervention trials suggest that beta-carotene is particularly promising with respect to cancer prevention. A high dietary intake of vegetables and fruits rich in beta-carotene is associated with a decreased risk of epithelial cell cancer. Beta-carotene is known to deactivate free radicals and excited oxygen, both of which have been implicated in cancer, so there is logic in this potential use of the micro-nutrient. There are now ongoing trials to evaluate the potential value of beta-carotene in cancer prevention. These include its effect, along with vitamin A, in those people occupationally exposed to asbestos and hence more prone to lung cancer. Others are looking at possible benefits of supplementary beta-carotene, vitamin A and vitamin E in reducing lung cancer amongst cigarette smokers. Cancers of the oesophagus, colon and skin are also being studied.

The ability of th body to resist cancer depends upon the status of its immune system and studies such as these may well confirm that this in turn is dependent on nutrition and in particular food factors such as beta-carotene, vitamins A, C and E, zinc and selenium. This equally applies to viruses and allergies, as our means of defence against all these conditions is a strong and healthy immune system. Adequate levels of micro-nutrients appear to be important criteria in maintaining a healthy immune system and the evidence presented in this book confirms it. Dr Jan de Vries is an eminent practitioner who combines a professional scientist's training with a clear, open and analytical mind that can assess the impact of nutrients and micro-nutrients on his patients' conditions. Here, then, is the ideal combination for an author who can write about viruses, allergies and the immune system, using his vast knowledge and experience of case histories to validate what is generally suspected about the needs of the immune system.

It is gratifying to see that the medical profession is at last admitting that prevention is better than cure and this is reflected in the studies mentioned above on cancer and in the dietary approach to reducing the chances of heart disease, for example. Those who read this book will see how far-reaching is the contribution that micro-nutrients give in protecting us against other manifestation of civilisation that are playing an even greater role in determining how healthy we are. The advice given by Dr Jan de Vries can only result in a better quality of life for all of us.

Leonard Mervyn,
BSc, PhD, C Chem, FRSC

1

Environmental Pollution

ACCORDING TO SEVERAL dictionaries, the definition of the verb "to pollute" is "to make unclean or impure, to corrupt or defile". A further definition of "pollution" is given as the "contamination of human elements and plant life by a harmful or offensive substance".

I felt prompted to look up the correct definition of the word "pollution" after a farmer and his wife visited my surgery on a beautiful sunny morning. As soon as they entered my consulting room I could see that they were both unwell. Chinese facial diagnosis has taught me to study people's faces for tell-tale signs as to their physical condition. This often serves as a pointer in my eventual diagnosis of their malady. On this occasion I noticed that the lady's eyes were sunken and encircled by black rings, but my immediate concern was centred on her husband. To me, he presented all the symptoms of Parkinson's Disease.

After they had seated themselves I observed them closely during our conversation, while concentrating on their story. I paid great attention to their answers to my questions and noticed that the word "pollution" cropped up several times.

Sometimes I wonder if mankind has reached the point of no return on the slippery path to self-destruction. As a naturopathic practitioner I feel the urge to remain close to nature and in harmony with creation. It saddens me when I see patients who had once enjoyed a happy life and good health, living at peace with themselves and the world, whose harmony is then disturbed by an outside factor. This change is often the result of artificial or man-made substances that interfere with the body's ability to work or function properly.

In this day and age, pollution whether it be on a large or small scale, deserves our full attention as our immune system is already subject to severe pressure. It deserves full attention not only from the medical establishment, but also from the world's governments if our fundamental human right to a healthy environment is to be protected.

The innate resilience of the immune system could be seen in due course in the case of the farmer's wife. With the help of several natural remedies I prescribed for her, she soon improved considerably, although the after-effects on her general health of industrial waste pollution will still be with her for some time.

Nature really is wonderful and, unless there has been irreversible damage, one can often witness a remarkable recovery when the correct measures are taken. My instinctive reaction to the husband had been that he could well be suffering from Parkinson's Disease, and indeed, I was not far from the truth. When I asked for a description of their problems I heard that they had experienced some extraordinary symptoms. They had also noticed a change in the behaviour of their cattle.

I recognised many of the symptoms as being similar to those experienced by others who had been affected as a result of using pesticides and insecticides, or living near to a site for industrial waste disposal. The list of such symptoms seems endless:

—tiredness
—headaches
—sore muscles
—painful and cracking joints
—nail discolouration
—pains in fingers and toes
—numb hands
—nosebleeds
—swollen glands
—urine changes
—pains in hips, back and arms
—itching eyes and ears
—cold sweats
—susceptibility to colds
—red, burning and swollen eyes

The farmer and his wife had also experienced a burning, peppery taste affecting the chest and tongue and sometimes a chemical taste in the mouth. The taste of tea, coffee or water had become abhorrent and this definite change in taste always became more pronounced after working with the cows.

While considering the condition of this gentleman, I remembered a recent article in the *Lancet* (of 15 November 1986) regarding a possible relationship between Parkinson's Disease and pesticides. The authors stated in the article that they had noticed a similar pattern in one of their patients and had reached the conclusion that they were dealing with an early onset of the irreversible Parkinson's Disease. This was very similar to the diagnosis I reached on my patient.

Since then I have studied many other cases which showed the onset of similar symptoms and I have often discovered a common factor, frequently referred to as pollution poisoning, or by a variety of similar names.

It is known that drinking water containing lead or aluminium can lead to degenerative diseases such as Multiple Sclerosis or Alzheimer's Disease; however, it is equally likely that these are triggered by some of the man-made and unnatural products we work with. It is possible that through exposure to such products one's health will suffer to a considerable extent.

The Institute for Toxicology in Mainz, Germany, claimed in a recent case to have found a link between the incineration at an industrial waste-disposal plant and a deterioration in the general health of residents in the surrounding area. This conclusion was based on the evidence of residents in the areas where such waste material was treated and the Institute did not hesitate to state categorically that it will affect the health of humans and cattle at random.

It is always sad to see a calf born with two heads. It is many times worse to see babies born with physical abnormalities and universal evidence to this end is such that it should send all alarm bells ringing.

In my patient's case a waste-disposal plant had been sited near his farm a few years previously. It was not long before his family's health noticeably deteriorated and the condition of his cattle also seemed to become progressively worse.

It seemed a foregone conclusion that the toxic fumes known as dioxins must have been at least partly responsible here. Black smoke constantly billowed out of the chimneys of this disposal plant and subsequently settled on the surrounding landscape. Consider, then, what possible effects this could have had on the milk produced by his cattle!

Prof. Dr Samuel Epstein from the USA raised a related point in a speech given at an Indiana conference. He pointed out that the surface waters are becoming contaminated; in many parts of the country these contain high levels of toxic chemicals and pesticides from agricultural run-offs. The surface water and therefore the ground water is becoming more and more polluted. The National Research Council of the Academy of Science, in a very conservative report, has estimated that pesticide residues at maximum legal limits are responsible for in excess of 1.5 million cancer deaths.

In many cases our food also contains high levels of pollutants, as does the air we breathe, the water we drink and the soil in which we grow our food. Atmospherically, we are constantly under the threat of pollution, which has become a formidable monster since the Industrial Revolution.

Toxic organic chemicals are now widely used. It is often overlooked or unknown that these could be carcinogenic. Other possible effects of these chemicals are miscarriages, infant mortality, nerve damage, immunological damage, lung, kidney or liver damage, or interference with many of the vital functions of the body.

Prof. Epstein, who has twenty-five years' experience in the field of toxicology and epidemiology, was quite frank in his warning. We all ought to be aware of our own responsibility.

We cannot afford to sit back and think that it will all sort itself out. It must be a concerted action. The following statement struck home with me: "If we act locally and think globally, we may possibly see an improvement in the present situation and be able to exercise some control." We must endeavour to think more naturally and live more naturally in a healthy environment.

The World Health Organisation's aim is Health for All by the year 2000. Its top priority should be to protect people's health with regard to the subject in hand.

I understand how difficult it can be where industry and one's livelihood is involved. But if we consider what is happening to our beautiful nature due to the loss of trees because of acid rain, and the resulting ecological problems and changes, we are then bound to realise that it is not merely our livelihood, but more our lives that are in danger.

It is my sincere wish that in the years to come this sentiment will find a response in an overwhelming universal political effort to regain a better climate for life on this planet.

Let us consider for a moment how we can be affected by the atmosphere through one of the busiest thoroughfares of the body. Through the mouth and down the throat we consume 40 tonnes of food in a lifetime, but we inhale near enough 500,000 yds^3 (380,000 m^3) of air. Particles ejected in a forceful sneeze have been measured at 103.6 miles (165.7 km) per hour.

We cannot afford to be lethargic about our responsibility when we read, for example, that Switzerland intends to send to Britain for disposal 28 tonnes of toxic filtered dust, which a Zurich firm had at first illegally exported to West Germany (reported in the *Glasgow Herald*, 26 March 1987).

We regularly hear and read about requests for permission to build yet another waste-disposal plant for radioactive material. One little breath could cause a whole lot of health problems and misery. It is of little or no use to receive financial compensation once the damage is done. Money is relatively cheap and cannot compensate adequately for the loss of health or life. According to the newspapers, in just such an area in Wales, twenty-three babies were born with eye problems. Isn't it time indeed that the government took action and had a long and hard look into these particular problems?

Forestry experts in most European countries have sounded a warning as to the disastrous conditions in their national forests, which are slowly but surely dying. The British government remains remarkably complacent, whilst other European governments now openly acknowledge these particular problems.

The recently originated name "acid rain" now has a familiar ring to most of us. It describes a complex of ingredients such as nitrogen oxides, sulphur dioxides and other harmful substances transported by the wind and deposited elsewhere through rainfall. Some of the most beautiful forests in the world are dying and it is said that in Switzerland alone in 1984 some 12 million trees, covering 14 per cent of its entire forest areas, had to be felled. In the Netherlands it is estimated that 40 per cent of its forests show signs of damage and more than two million acres (800,000 hectares) of forest in Czechoslovakia are threatened, whilst half a million acres (200,000 hectares) of forest in Austria are affected (reported in *The Observer*, 19 October 1986).

The damage to our rivers and lakes is already extensive. Even in the sparsely populated areas of Scotland where there is no industry to speak of, testing some of the lochs has shown that some white fish are now turning blue and are ailing. What are we doing to our natural environment? Why has mankind allowed itself to be led along this path of self-destruction?

It is imperative that we act quickly. Our wildlife is suffering and with the threats hanging over our food, water and air supply, it is absolutely essential that the damage that has already been done is brought to a halt and stricter rules are laid down and adhered to.

At a rather impressionable age I read the book *The Dance with the Devil* by Günther Schwab and soon realised the extent of such problems even then. Some of the facts stated in that book I still vividly remember.

One of the statistics which greatly impressed and worried me was that the factory chimneys of the American steel centre every year emit seven million tons into the air of carbon dioxide from coal. I also remember reading that the carbon monoxide emitted into the air in Germany by the large numbers of cars on the road had been proved to cause health problems.

Against much lobbying and opposition the British authorities decided some years ago to legalise the addition of fluoride to our tap-water supplies, as it was claimed to be beneficial to our teeth. This measure is now forced on innocent people who regularly absorb these unnatural fluoride cells, unaware that they could well influence their health in other ways. In a vein of cynicism I would ask the Minister for Health why not add a laxative to our water supplies, as half the British population suffers from constipation? Morally, we cannot act on behalf of people without putting all the facts on the table. Failure to do so is considered an infringement of human rights.

So many years ago Günther Schwab already had the foresight to warn us against the threats of pollution and pointed out its complications. He warned us against artificial fertilisers and the use of pesticides and insecticides, even though he was already aware of the tremendous growth in the world population and the necessity to increase food production.

Chemical and often poisonous pesticides and insecticides have caused universal damage. The figures for the use of pesticides and insecticides are astronomical; billions of tonnes are applied in the worldwide production of food. The users should be made aware that by applying these artificial substances the good as well as the bad germs are killed off. At all times we should endeavour to keep our food as natural as possible in order to give us the energy we need in our daily lives.

The *Daily Record* of 9 June 1988 featured an article about the fears of an increased cancer incidence in the area where the Scottish Dounreay nuclear plant is situated. The report dealt with the incidence of leukaemia among local children. An independent investigation revealed that the rate of blood cancer among children in nearby Thurso was twice the national average. In my book *Cancer and Leukaemia* I have pointed out how disconcerting similar statistics are in several parts of the world where atomic plants have been built.

Such information is often treated with disparagement by the authorities, as it does not fit in with their future plans. This, however, should not detract from the fact that we must realise that we are born from nature, that we belong to nature and that if we obey the laws of nature we obey the laws of God. Fortunately, more detailed studies are continuously taking place and the pressure on governments is greater than ever before.

The edition of *The Scotsman* of Friday, 27 June 1988, featured an article with new evidence relating to a "cocktail" of chemicals found in recent soil samples. This casts serious doubts on previous findings indicating that health problems among humans and animals could not be linked to industrial pollution.

Hardly a day seems to pass without one of the daily newspapers featuring some item or other on the subject of pollution. Pollution poses a very real threat to our existence and to that of animal and plant life. It is up to us how we counteract this and protect ourselves. If ever natural cures were prominent, then it must certainly be today, because effective remedies to safeguard ourselves and at the same time boost our immune system are within our grasp.

Another book that impressed me greatly when I read it some time ago was *Civilised Man's Eight Deadly Sins* by Konrad Lorenz. The author relayed his message well and in his summary he points the finger at today's problems and lists the factors as follows:

1. overpopulation of the earth;
2. devastation of our natural environment;
3. man's race against himself;
4. the waning of all strong feelings and emotions caused by self-indulgence;
5. genetic decay;
6. the break in tradition;
7. the increased indoctrinability of mankind;
8. the arming of mankind with nuclear weapons.

From this book I would like to quote the following passage:

> The processes of dehumanisation as discussed, give support to the pseudo-democratic doctrine which maintains that the social and moral behaviour of man is in no way determined by the phylogenetically evolved organisation of his nervous system and his sense organs, but rather that this behaviour is determined solely by the condition to which in the course of his ontogenesis he is exposed by his particular cultural environment.

I am well aware that many other books and articles have appeared on the subject of pollution. I am also aware that many people share my concern. As a practitioner, supporter and advocate of naturopathy, I call upon everyone to take up the fight against the increase in pollution. A sensible solution to the disposal of waste material must be found and we must develop a satisfactory approach to agriculture that is more in line with the laws of nature. Furthermore, the victims of pollution need assistance to rebuild their immune system. Plants and animals also deserve our consideration in this matter.

What it basically boils down to is that the harmony between man and his Creator must be restored. Occasionally we experience that wonderful feeling of being in harmony with God's nature. Let us therefore endeavour to reach our common goal by working together towards a better awareness of what is good for mankind.

2

Chemical Waste

A GOOD WHILE AGO I was consulted by a pleasant young mother. My first impression was that she was in inner turmoil and most certainly she conveyed without words a nervous disturbance that alerted me. She told me that although she was happily married and had no financial worries to speak of she sometimes experienced such anxiety that at times she felt suicidal.

Sometimes I compare my job to that of being a detective, because by probing and questioning the patient I find the root of the trouble. In this case, however, I did not get very far. There was something worrying this patient, but the cause of the problem eluded me. Nevertheless I was greatly concerned at the effect it was having on her.

I prescribed some general treatment to ease this patient's nervous tension and that seemed to help her to a certain extent. However, I cannot say that I was happy with the slight improvement we achieved. On one of her subsequent visits she was accompanied by her husband, who happened to remark on the coincidence that the problems had only started to appear after they had moved house. I let this sink in for a minute and wondered if this might just be a clue. I further enquired as to the circumstances of the move. The young woman loved her new home and told me that she felt settled in the new environment. Then I learned, quite by chance, that she now lived not far from the site of a chemical factory where a number of well-known poisonous chemicals are manufactured. As a result of this information I decided to do a blood test and found the presence of a miasma, which unfortunately I was unable to exactly define.

For the time being we struggled on with some homoeopathic antidotes and she lethargically began to accept the lower quality of life she was experiencing at that time. Eventually, after long consideration, they decided to move house again. In my mind there was little doubt about a connection between her nervous anxiety and her living environment. I had previously come across a similar case where the patient had been affected not only mentally but also physically.

I had not seen her for quite some time when she and her husband visited me out of the blue. I hardly recognised this very attractive and confident person as the patient who had previously attended our clinic. However, I could not take credit for helping her to regain her health and confidence. She had flourished in her new surroundings, away from the chemical influences, and we were not left in any doubt that these had certainly undermined her health. The results of the move could not be disputed and we had to accept that the change was there to be seen after other measures had failed. Moving to a new environment seemed to have brought about this change for the better.

Sceptics may claim that her improvement could be attributed solely to the new house and environment, but then we should not forget that she had initially felt happy in her previous surroundings and had been loath to move away from there. It was not a question that the new experience had livened her up to the extent that she felt much better. I am perfectly convinced that the change of living surroundings — away from the chemical plant — had been the answer to her problems.

This incident reminded me of an article I once read in one of the daily newspapers about how dioxin had slowly been polluting a neighbourhood during the early seventies, and how this had considerably undermined the health of the local population without them realising. Another article in *The Sunday Times* (24 December 1984) reported how dioxin had caused alarm in a village because of the increase in facial deformities among newborn babies.

In an article published in the *Sunday Telegraph* magazine (15 March 1987), my friend Barbara Griggs discussed with two doctors the toxic threat that could affect even the next generation, and warned that each year babies face an even stiffer challenge to their physical constitution than was faced by older children in their day. Chemicals present in their food and environment cause sudden outbreaks of violent behaviour and very often pain in certain children.

The Asthma Research Council estimates that one child in ten is asthmatic, which is the equivalent of three children in an average primary school classroom. Fortunately, many of these childhood asthma symptoms disappear by the time the child reaches its teens, but by no means do all children outgrow these problems. Deaths due to asthma are rising steadily in the United Kingdom. In 1983 the official figure given was 1,863 people, while in 1985 that figure had risen to 2,167. Figures for eczema have also shot up, and complaints about hyperactivity in children, funny tummies, wheezing, odd rashes, headaches, etc. rise daily.

The extent to which chemical pollution can ravage people's health is still not clear. A study published in the *Lancet* suggested that our bodies cannot get rid of toxic chemical contaminants by natural secretion only. Much more stringent measures are needed to destroy the effects of contamination. The immune system must therefore be encouraged by building up one's natural reserves and this is more important than ever before. In a later chapter I will give more details on how we can boost our immune system in order to protect ourselves.

Pollutants or chemical-waste materials can and often will destroy that vital immunity in our lives. What we are learning now, however, is that these problems do not only occur in areas of which we have long been suspicious, such as contamination of food or water; but atmospherical contamination has also now become a real threat.

Peter Mansfield and Jane Munro have written an excellent book called *Chemical Children*, which explains in detail the medical repercussions which can result from pollution or incorrect disposal of chemical waste. Reading this book, I cannot fail to remember the basic naturopathic principles that improper foods affect the stomach, the spleen and also the brain; air affects the energy in the lungs, the large intestines and the skin and sunlight influences the energy of the heart pericardium, the brain and the skin. Deficient food will also affect the liver and gallbladder functions.

In naturopathy much emphasis is placed on water, air and sun. This form of treatment has been appropriate for centuries, but has become almost impossible today. New methods for our protection will have to be devised as there are few instances left to treat ourselves according to the original dogma of naturopathy, because of interference by mankind, specifically since the Industrial Revolution.

I remember the case of a young hyperactive boy who was brought to our clinic. I discovered that apart from being allowed to eat the wrong foods he had also been in contact with chemicals. Today he is a normal teenager who applies himself to his studies and enjoys his sport, and I am sure that this is because the disturbance in the balance between mind and body has been restored. The youngster was a victim of chemical pollution and his parents could not understand what was happening to their son. The doctors who had been consulted referred to his problem as a "mystery illness". The parents looked for help everywhere and it was a while before we discovered that it was caused by a specific chemical that had been used in their house to treat woodworm.

On 10 June 1984 the *Sunday Mail* published an article on two babies: a one-year-old baby girl who was mentally retarded and suffered from epileptic fits, and another baby blind in one eye since birth. Oddly enough, these births occurred in the vicinity of a chemical plant. Although such instances are very difficult to assess and we would rightly be accused of scaremongering and of jumping the gun if we assumed a direct connection, why this should be so remains a big mystery. The particular areas where such occurrences are more common certainly need thorough investigation.

In the summer of 1973 it was discovered that a chemical company had made a mistake in fulfilling an order. Instead of magnesium oxide – a harmless antacid — a truckload of highly toxic industrial chemical polybromenated biphenyls was delivered to a large farm and animal-feed supplier. As the two substances are similar in appearance the error went unnoticed. Then cattle started to fall ill; some died, while others showed malformations. Also the incidence of still-born calves increased alarmingly.

When I read about this I could not help wondering what happened to the consumers of the milk produced by the contaminated cattle. Do we know what happens to the people who drink the milk of cattle fed on grass that has very likely been fertilised with chemicals? How much do we really know about the possible far-reaching effects of chemical fertilisers?

In point 10 of the 245-T dossier of the Union of Agricultural and Allied Workers we are reassured that the experts are aware of the risks of 245-T. We are informed that if it is handled properly and if the spray is used correctly there should not be any danger. This just leaves me to wonder who decides when conditions are "suitable" to spray the crop with this fertiliser and, moreover, whether any account is taken of sudden changes in weather conditions.

Point 11 queries the remark made by one of the advisers as to the supposed safety of 245-T, based on the fact that he used it in his own garden. This remark deserves at best to be regarded as facetious. There is no doubt in my mind that too many of these toxic substances are used and often by people unaware of the disastrous possibilities that could follow in their wake.

The committee readily admitted that they were still not clear about the real effects on human beings of 245-T and its incredibly toxic contaminant dioxin. This to me suggests an alarming level of ignorance. From its own report I quote: "It shows once more that one has to be very careful."

The *Daily Express* of Friday 8 July 1988 featured an article on a meeting of the British Medical Association where it was claimed that "pesticide contamination of food may be causing allergies and cancer". It was alleged that of 171 patients attending an allergy clinic, 46 per cent were shown to be sensitive to pesticides. Figures show that last year over 26 million kilogrammes of pure pesticides were used in the United Kingdom. The article went on to say: "All of us who eat horticultural and agricultural products are taking in chemicals shown to be connected with cancer."

On a visit to a farm during my pharmaceutical studies, I once saw the farmer injecting a cow with such a high dosage of penicillin that I was greatly worried at the thought of the milk produced by that cow reaching some unsuspecting consumer. Fortunately, restrictions in this field have since been introduced, yet there still remains far too much freedom, especially concerning the use of chemicals.

Sometimes I sympathise with the golfers on a golf course who enjoy the freedom of nature and its beauty. Do they ever stop to consider by what means the grass on the course is kept so beautifully green and in such good condition and what they could be inhaling with every breath they take? Remember, the mileage of breath we inhale every day, as mentioned in the previous chapter.

The female patient mentioned at the beginning of this chapter had been unwittingly affected by fumes from the chemical plant in the vicinity of her home. It is interesting to note that many people who live near toxic plants often feel sickly, not necessarily because of the nauseating smell, but also because of the fumes that are emitted. The body is a wonderful indicator of what is not agreeable to us and it is a very natural reaction to feel sickly if an obnoxious substance is to be rejected. Pay attention to these alarm signals.

It is a known fact that in certain parts of the United Kingdom people experience a sickly feeling after drinking tap-water. This, to my mind, deserves further investigation. So far the regional councils have not managed to pinpoint what causes this and the only explanation as yet is that it must occur naturally as a result of decaying organic matter.

What happens to the toxic waste material from the manufacturers of plastics, lubricants and electrical equipment? In the manufacture of their products a great deal of toxic material is produced, but are the guidelines laid down for disposal of waste products stringent enough and are they fully adhered to? Do the benefits of living in a modern society outweigh the disadvantages?

3

Viruses and the Post-Viral Syndrome

WHENEVER I CONSIDER the ever-growing list of patients who attend our clinic because of viral problems, I cannot help but wonder about a connection with the environmental factors discussed in earlier chapters. It seems to me that the post-viral syndrome or Myalgic Encephalomyelitis (ME) is near enough reaching epidemic proportions in Scotland.

Undoubtedly the efficacy of the immune system is subject to the three forms of energy we need to sustain life, i.e. food, water and air, and therefore we are constantly endangered by infections or viral invasions.

I recall a public debate in which I took part together with a highly qualified and influential member of the conventional medical establishment. He was indeed a very intelligent and clever person, who carried a high level of responsibility in medical circles, but he had the reputation of relating everything to percentages, figures and data. Worse was that he made no secret of the fact that he was averse to any form of alternative therapy. Our debate lasted for one and a half hours and we had the pleasure of parting as the best of friends. His parting shot was that we should collaborate much more as, after all, our mutual aim was to alleviate human suffering.

One of the points I raised, which is appropriate in this chapter, was that although orthodox medicine has given us ample scientific evidence and data, why then have we not made more progress in the field of viruses and allergies? Our hygienic methods and knowledge would seem more effective than ever, yet these problems are still increasing rapidly. Despite ever-increasing knowledge in so many fields, we do not seem to make any inroads into solving the present viral and allergic problems. Moreover, from Eastern countries, we hear more and more about the dangers of invading parasites.

When I asked him for suggestions on how to deal with the post-viral syndrome or ME, he used the often-heard platitude that it would very likely burn itself out and in the meantime we would have to learn to live with it. Knowing how many people are handicapped by this relatively modern and growing phenomenon and even seeing some of them being subjected to psychiatric treatment as a result, I cannot possibly sit back and wait for it to burn itself out.

Where have we gone wrong if we cannot find an answer to these particular problems, while at the same time realising that more and more people seem to fall prey to allergic reactions? Obviously we have failed somewhere — but where? Even though conventional medicine rejects this line of reasoning, I wonder if it could not stem from the way we live at present, from our modern-day dietary habits or atmospherical influences.

It would be impossible to deal with all of the presently known viruses and allergies and I intend only to select a few for discussion in this chapter. Let us realise, however, that if we choose to sit back and let ME burn itself out, more victims will be claimed. We must do something and tackle the problem from a sensible angle and use the as yet limited information and knowledge available to us.

I am pleased that among my patients with allergic or viral problems I can also count practitioners trained in the conventional medical methods. I am even more pleased that I have been able to help most of them and the progress monitored has been a joint effort. Seeing the growing misery I am prepared to consider any advice or opportunity to try and help my patients who suffer from these debilitating conditions.

During our medical studies we are taught about serious viruses, less-serious viruses, aggressive viruses, and slow-working viruses. The medical textbooks contain difficult names such as Coxsackie virus, or e.c.h.o. virus (enteric cytopathogenic human orphan virus), both of which, by the way, belong to the family of the enterovirus. However, it is certain that all these viruses, disregarding the complexity of their names, could lead to serious problems.

Names such as AIDS, or the hepatitis virus, or the Coxsackie virus, unfortunately no longer have a totally unfamiliar ring about them. Due to an increase in their incidence these names appear more and more regularly in the press.

The person affected by the Coxsackie virus will most likely be diagnosed as suffering from the post-viral fatigue syndrome. I am happy to agree with certain investigations which claim that not everyone who is suspected of suffering from ME actually does so. Yet I also know that there are people who suffer unknowingly from this complaint and have done so for a number of years. Often the gradual deterioration in their health has cost them dearly in their private life and/or their standing in society.

An acute viral hepatitis is usually noticed fairly quickly and often finally brought under control. It is, however, much more difficult when a slow virus or multiple viruses undermine one's health. I have seen it with one of my patients where multiple Coxsackie viruses played an enormous role in a post-viral fatigue syndrome.

In patients who have been unwell for quite some time I have discovered the Epstein-Barr virus, belonging to the herpes family. Sometimes we recognise infectious conditions or interaction of viral infections due to which the immune system has been attacked. In such cases we occasionally observe enzyme abnormalities. It is not only with post-viral syndrome but also with post-viral neurological illnesses that these problems only come to light after extensive tests. Sometimes these tests show the immunological results as a severe lymphocyte dysfunction, something I have already discussed in my book *Cancer and Leukaemia*.

Very little was known about Myalgic Encephalomyelitis or ME. This condition affects the brain, the central nervous system and the muscular system. It used to be largely ignored and patients were informed that it was all in the mind, which unfortunately still happens sometimes today. ME seems to have gained a considerable foothold in certain areas nowadays, although elsewhere it only occurs sporadically. Glandular fever presents us with a similar patchy pattern.

The disease was first noticed in Los Angeles in 1934 and was after that recognised in Iceland in 1950. It became known as the Royal Free Disease in the United Kingdom in 1955 and has also been virulent in other parts of the world. It seems rather odd that it was originally considered to be more prevalent among females, yet nowadays it seems to affect females and males equally ruthlessly.

Myalgic Encephalomyelitis may manifest itself through different symptoms, which sometimes makes it difficult to reach an early diagnosis. Complaints may range from vomiting and diarrhoea to sore throats with coughs and colds, while later patients will complain of lethargy, weakness and exhaustion. Often they will volunteer that they have never felt so ill in their lives. In our practice we deal with many patients' back and neck problems, and I have been told by ME patients about severe cramping pains in the neck and shoulders, arms, chest and stomach. Other patients report pins and needles, depression, a constant need for more sleep, shivering attacks, muscle twitching, funny feelings in the back of the throat, etc. Thus you can see how widely varied the symptoms can be.

Unfortunately, it often happens that no timely action is taken or help is sought and therefore months, or even years, of suffering are experienced. The duration of the illness varies greatly, but it leaves the immune system so greatly impaired that while recovering from one virus it is not unknown to fall victim to another virus as a result of the decrease in one's resistance. Then the whole sorry story may repeat itself.

One such case was that of a successful businessman. After having attended a cocktail party he felt unwell on his way home. He was perspiring profusely, felt dizzy and had a sore head. Within a few weeks' time he was diagnosed at the hospital as suffering from the post-viral fatigue syndrome. Tests revealed that he had been attacked by several cocci. He followed his specialist's advice, but did not experience any improvement. He then came to our clinic and after a long discussion he agreed to some changes in his diet and I prescribed several remedies for him:

—large doses of Oil of Evening Primrose;
—Harpagophytum — a herbal remedy from the Vogel range;
—a remedy to boost his immune system, which is produced according to my own formula.

With these three remedies he overcame most of the symptoms and felt very much better. I felt very sorry for him indeed and I will repeat here what he told me:

"No one who has not experienced this first-hand can possibly have any idea what a person with ME really feels. One feels like dying and absolutely fatigued all the time. Every day it is a real struggle to turn up for work."

During several tests on another young patient it became apparent that apart from having been affected by several viruses and *Candida albicans,* he had also been in contact with the pesticide 245-T, mentioned earlier. I did feel that this diagnosis vindicated my belief that the immune system is very much involved in these problems. I imagine that the immune system itself will perhaps fight and try to deal with infections, but it has to work extra hard when it is attacked by such viruses.

I have already mentioned the difficulties in diagnosing ME due to the variety of symptoms. However, it can usually be diagnosed more clearly with an iridology test than with a blood test, which may show evidence of a chronic virus. Furthermore, the ME problem will affect people much more if they are under physical stress or if they are suffering from another infection. Smoking and/or drinking accentuates the symptoms, as does emotional trauma. I would imagine that this last factor could possibly be the reason that this problem is sometimes diagnosed as a mental condition.

A detailed questionnaire completed by several hundred members of a self-help group resulted in the following conclusions being obtained:

—65 per cent of the sufferers were female: 35 per cent male;
—83 per cent of the sufferers were over thirty years old and the forty-fifty age group was most easily affected;
—72 per cent had been diagnosed as suffering from the Coxsackie virus;
—26 per cent had been tested for the above, but proved negative;
—72 per cent were below a 60 per cent level of recovery, with 12.9 per cent showing no improvement and 10.2 per cent being worse than at the onset of the illness;
—2.7 per cent indicated a full recovery;
—43 per cent had either family or friends who also suffered this affliction;
—20 per cent had taken early retirement or had lost their jobs because of the illness;
—3,500 working weeks had been lost (and housewives were excluded from this figure as they classed themselves as unemployed). The loss in man hours represented a total of somewhere in the region of £400,000 for allowances and benefits.

This survey was conducted in 1987 and the figures obtained have most certainly been overtaken now, as ME still appears to be a fast-growing problem.

Extra care must be taken if the ME patient is prone to respiratory or gastro-intestinal problems. In such cases I often consider prescribing germanium, a mineral which has proved invaluable, especially with pain and breathing problems. In the United Kingdom this remedy is marketed by the company Nature's Best.

In our clinic we occasionally used to employ a young and intelligent girl to help out as a holiday relief worker. She was studying to become a doctor and we did not see her for quite some time because of her studies. Finally, she came to consult me as a patient. She had become stuck with her studies for inexplicable reasons and even though her mother was very wary about alternative treatment they had explored every other avenue without any success. I have never seen such a radical improvement in any patient. We changed her diet, introduced a vitamin supplement and prescribed a high dosage of the mineral germanium. Within two months she was nearly back to normal and able to restart her studies. Since then she has overcome her problems totally.

Her hospital consultant confirmed that even her brain had been affected, but thanks to the four-dimensional mineral germanium she completely recovered. She had followed my instructions on diet and remedies to the letter and her mother fully supported her. This positive commitment proved to be the answer in her case.

All too often I see the rate of a patient's progress slowed down because of lack of co-operation. An all-round sportsman whom I greatly admire progressed well until he overlooked his dietary instructions while on holiday. During that relatively short time his achievements to that date were very largely undone. Unfortunately, in such cases the patient generally becomes more and more depressed and at the same time unable or unwilling to admit that they themselves are to blame. This see-saw movement in their progress sometimes causes a disturbance in their sleeping pattern and balance, blurred vision and low body temperature. These symptoms in turn will make the patient yet more depressed, which certainly does not help to solve the problem.

It all comes back to obedience to and respect for the laws of nature. Although these may appear too simple to be effective in our sophisticated world, we can rest assured that nature will take its own course and will not accept interference from mankind if it cannot cope with it.

Myalgic Encephalomyelitis is sometimes called the "mysterious disease" and it is difficult to advise on a general diet.

In my experience, patients who may have allergic reactions or a *Candida albicans* virus will very likely need a different diet to others. Certain foods are often suspect and generally I recommend that fermented products, cheese, wine, mushrooms and chocolate are avoided. However, for other patients the requirements may be different. On the whole it is quite safe to follow the *Candida albicans* diet, which has generally been of help to patients. Even then I find that the products mentioned above often exacerbate the problems. In order to keep the lactobacillus situation in check a daily tub of natural yoghurt can also be very useful.

Once again we must ask ourselves how much influence have the three forms of energy on the ever-increasing problem of viruses, allergies and post-viral syndrome? Do we know if what we eat and drink are body-breakers or body-builders? What role, if any, is played by atmospherical influences on our food and water supply, e.g milk and other products obtained with the aid of artificial fertilisers?

I consider it very disturbing that in Australia the suspicion has arisen that outbreaks of ME may be associated with people who have been in contact with certain agricultural pesticides. It is thought possible that the ME syndrome may be due to a disturbance in the body's normal immune response, caused either by viral agents or by toxic organical chemicals. The illness seems to be characterised by fluctuating symptoms and relapses of malaise and flu-like attacks. A considerable number of ME sufferers also have allergies, especially to common foodstuffs and chemical additives.

From routine blood tests it is practically impossible to pinpoint what is happening. However, abnormalities do show up in the red blood cells and a gross disturbance of the immuno-regulation affecting the body's cellular immunity. This seems mainly due to abnormalities in the lymphocyte system, where ME patients are more susceptible to infection than other individuals. Abnormal muscle-cell metabolism and enzyme deficiencies are also often noted. It is therefore of the greatest need that sufferers of ME or any virus should try and eat organically grown food in order to fight off infections and to condition a good immune response.

This problem can be severely limiting on our actions, especially for active sportspeople. One of Scotland's well-known footballers, David Provan, knows this only too well. I have known him for years and although he has done everything to overcome his illness, he has been forced to retire prematurely from competitive football. Luckily, he still is involved with his much-loved sport and he is able to coach youngsters. If the problem can be diagnosed and arrested in the early stages, the chance of complete recovery is obviously much better. Immediate action is necessary, as is a sensible diet and some remedies to boost the immune system. These I will discuss further in a later chapter.

Thousands of viruses exist and these include the smallest known disease-producing agents. A good healthy immune system is essential, as is rest, sleep, as well as dietary and medical care. Viruses use different ways to enter the human body, of which the most common is through the respiratory tract. Here we often see that people with existing respiratory problems could suffer dramatically unless immediate action is taken.

Recent research has indicated the existence of the RNA virus particle inside the muscle cell. The RNA multiplies itself and if the virus leaves the cell it can initiate a process of abnormality. When using the enzyme therapy for severe cases of viral infections, in an effort to boost the immune system, I have sometimes seen a sudden change for the better. A breakthrough can be effected by forcing the cell into a build-up of extra enzymes together with increasing vitamin levels through the use of supplementary vitamins, minerals and trace elements such as selenium, magnesium and zinc and by keeping an eye on the amino acid supply.

It is now generally accepted that the essential fatty acids play an important part in this process and if ever the seeds of the Evening Primrose, sometimes called Omega 6, came into their own, it is in incidences as described here. The oil of these seeds helps to improve the concentration and relieve neurological symptoms. We have to explore every angle in our efforts to find an effective way of dealing with these problems. If old-fashioned methods prove useful, let us not be too proud or pedantic to use them in our search for relief from the invasion of viruses. Volume 294 of the British Medical Journal, dated 7th February 1987, features an article that looks into the complexity of the post-viral syndrome or the Royal Free Disease. It states that as yet no certain cause has been established to this perplexing syndrome, but the application of gene probes and mono-clonal antibodies may provide one.

During a lecture at a teaching hospital a medical student asked me why I have so much faith in Oil of Evening Primrose capsules, as I had mentioned that I prescribed them for various purposes. He considered this to be "quackery". From his accent I realised that he was Scottish and asked him if he had ever heard of the poet Robert Burns. Of course he had, he declared condescendingly. I then enquired if he was familiar with the letter Burns had written to Dr Moore in which he expressed his opinion about the mistakes and blunders that were made which were due to ignorance. This statement still holds true today.

How ignorant we were when, nearly two decades ago, I first prescribed Oil of Evening Primrose. It did not take my patients long, however, to come to appreciate its medicinal value for various health problems, among them major problems such as viruses and the post-viral syndrome. When I prescribed it in those early days, I was sometimes laughed at, until eventually the medical establishment also came to realise the extraordinary powers of the Evening Primrose.

I pointed out to the medical student how short-sighted that attitude had been as at present Oil of Evening Primrose capsules are also regularly prescribed by the same hospital specialists who initially had been so sceptical about them. Today many recipients of this medication are willing to testify to its beneficial properties. A long time ago my great-grandmother was already prescribing Oil of Evening Primrose and it was already known in the old folklore of medicine. Unfortunately, for a period of time it became unfashionable and was forgotten. Sometimes it is possible that we look too strenuously for the solution to a problem, when the answer all along has been at our fingertips.

I will quote here a few lines from a poem written centuries ago on the Evening Primrose which says that it contains a life "unseen":

> And far from the world's infectious view,
> Thy little virtues safely blew,
> Go and in day's more dangerous hour,
> Guard thy emblematic flower.

The innate energy in the seeds of the Evening Primrose is of great benefit and of tremendous aid in the programme I use for the treatment of viral infections or for the post-viral syndrome.

While working on the preparation of this chapter I was happy to speak again to an ex-patient — a bank manager — who had contracted ME seven years ago and as a result had been forced into early retirement. He praised my positive attitude and perseverance and claimed that thanks to that tenacity he was now fully recovered and back at work.

It is impossible to force the situation and patience is required. Take sufficient rest and relaxation and never give up hope of finding out the exact nature of the problem. It could be an allergy or an underlying infection, or even *Candida albicans*, which in many cases also plays a role. This I will come back to in the next chapter.

Over the years I have obviously collected and compiled many case histories. However, I would like to close this chapter with a case history put together by a patient of mine, as I could not possibly portray his emotions as he can himself. He has spent a lot of time on his story and supplied all the relevant background information, which serves to give us an insight into the suffering ME patients must endure.

The story of a 23-year-old male ME patient

As a child I never had any serious illness. Being a fit and healthy lad I was very keen on sport. I was a regular member of the school football, cross-country and athletics teams, and was also a keen skier and swimmer.

I also found the time and energy to do some part-time jobs while at school. I worked as a vanboy, a milk boy and a sales assistant in a petrol station (not all at the same time, I hasten to add). On the whole I was a lucky lad.

Academically I was average. However, I got on very well with people, especially in my capacity as deputy head boy during my last year at school. I was involved in organising many sporting and social events. Knowing that I enjoyed this sort of activity, catering seemed like the obvious choice of career.

In 1983 I left school with eight O-grades and two Higher grades and started my studies in Glasgow in Hotel Catering and Institutional Management. It was a sandwich course which required many months spent working in industry. (I worked my placement at a hotel in Aviemore.)

I was forced to take a year out of college after failing an examination at the end of my second year. Therefore, instead of beginning my third year of college during October 1985, I remained at my industrial placement post at Aviemore.

At that time I was working at the hotel as a porter. My duties included:

> *Driving:* Collecting guests at the station, transporting staff, and any other tasks which involved transportation.
> *Function and conference set-ups:* This was a fairly physical task. It involved setting-up tables, chairs, staging, partitions, etc., and also clearing these away.
> *Carrying cases:* This was another job that involved lots of energy, particularly when five coaches would arrive at night, to leave again in the morning. (Why is it that American tourists' cases are twice as heavy as anyone else's?) I should also add that there were no lifts in the hotel.

Other jobs included shifting beds around and basically anything else the management could think of. If it needed doing, see the porter! Portering was actually one of the better jobs in the hotel, as the work was so varied and the money was a bit better than some other jobs in an hotel.

Then towards the end of autumn 1985 my life started to fall apart as a result of illness. At first the illness presented itself with anxiety-type symptoms which worsened over a period of months into fatigue so profound that I could hardly lift my arms above my head.

During the summer of 1986 a psychiatrist at a hospital in Paisley was convinced that my problems were the result of the post-viral fatigue syndrome. This was later confirmed in December 1986 by a consultant at a Glasgow hospital, after a muscle biopsy and blood tests.

My illness, which is now chronic, has run an unremitting course for more than two years and at present shows (for me) very little sign of improvement.

The progression from a very active life to what can now only be described as "non-life" is very hard to bear, at an age when one should be at peak fitness and enjoying all pleasures that such a treasure brings.

I have been forced to give up all sporting activities as well as my management studies, and at present I cannot even cope with part-time non-skilled work, which I have attempted on several occasions since becoming ill. The illness has left me with so many problems that I find it hard to relax and communicate effectively with other people.

However, the worst aspect of post-viral fatigue syndrome (or Myalgic Encephalomyelitis) is feeling so ill at times that you think you are going to collapse and die. Two and a half years on these awful "turns" still haunt me.

I will now try and describe as best I can how the illness began and progressed to its present-day state.

November 1985

Suffered flu-like symptoms and the general practitioner advised a few days' bed rest. I began to suffer from chest pains. I also felt generally run down and depressed. There were possibly other symptoms which I cannot remember because it was such a long time ago and my memory is pretty poor at the moment.

My main worry was that I had acquired some sort of heart condition. I consulted a local medical practitioner who assured me that I was physically sound and prescribed some anti-anxiety tablets.

December 1985

During this month my health began to deteriorate further. Chest pains became more frequent and severe, and I seemed to vomit regularly. At the time I associated the reason for the sickness directly to alcohol consumption. With hindsight I know that I was drinking far more than was good for me, perhaps this was because I had become very tense and nervous.

I began to take weak turns, when the energy just drained from my body and I thought I was going to collapse. I was sent home from work several times and a GP would be sent for. However, on examination nothing was found to be out of the ordinary. In fact, usually by the time the doctor arrived I felt a whole lot better, which left me feeling quite embarrassed.

Towards the middle of December 1985 I took very unwell while travelling to England to see a pop concert. Some of the symptoms I experienced were chest pains, acute dizziness, disorientation, trembling turns, exhaustion, sweating, breathing difficulties and blurred vision. Then there were the turns where I felt so ill and weak that I literally thought I was going to die. I had never felt so ghastly in my whole life.

I was also aware that I had problems with my waterworks. There was general discomfort and burning when passing urine as well as urethral discharge (later diagnosed as non-specific urethritis) and not related to the other problems I was having.

When the bus arrived in Liverpool for the concert I was praying very hard that I would not spoil the whole occasion for my friends by collapsing. I was very disorientated and being in a place which was unfamiliar to me, with thousands of people dashing around, was scary. I couldn't wait to get back on the bus and back to Scotland.

To cut a long story short, I made it back to Scotland, albeit in an exhausted state (but nevertheless relieved). Instead of going all the way up to Aviemore with my friends, I stopped over in Paisley at my parents' home as I badly needed bed rest as well as the reassurance that I was going to be OK.

The following day I completed my journey back to Aviemore and the first person I went to see was the doctor. As a result of my day's bed rest I was feeling a bit better. I was then more concerned about the problems I was having with my waterworks. Why? Because I was humiliated as I thought that I must have contracted a venereal disease.

The general practitioner referred me to a special clinic for my urethral problem where NSU was diagnosed and I received a two-week course of antibiotics. For my other problems the doctor prescribed a drug called Prothiaden which, after taking it, knocked me absolutely silly and caused violent vomiting. Needless to say I only took two or three tablets and threw the rest out.

Things were really gearing up for my worst Christmas to date. I was taking antibiotics for NSU, but I was certainly beginning to struggle at work. I felt very tense all the time and I was very worried because I knew my health was deteriorating. I wasn't in the same jovial Christmas spirit as everyone else, which was unusual for me.

Over the Christmas period I began to feel very unwell. I was sick regularly, with pains in my stomach and chest. I began to suffer from persistent headaches.

January 1986

I was by now persistently anxious and tense. I don't normally keep a diary, but I did keep a few notes at the beginning of the month which went something like this:

> I was sent to the GP's surgery in the middle of a shift at work because I felt weak, as if I was going to collapse. My blood pressure was found to be high and I was given painkillers for the severe headaches and told to go back for a blood test on the following Monday.
>
> 4 January 1986
>
> I was again sent home from work. Headache, general weakness (particularly in my legs) and spent all day in bed. Took the following day off work.

6 January 1986
Received blood test at surgery. Later the personnel manageress asked me into her office to discuss my problems. I was offered a change of job (from porter to barman) as well as two weeks' leave commencing after the busy period, i.e. from 14 January.

9 January 1986
I was sent home from work feeling unwell — weak, light-headed, dizzy, nauseous. I got the doctor to visit me who said that I was run down and also had a throat infection. He prescribed penicillin tablets (which I did not take).

N.B. The blood test proved negative.

It is probably true to say that these brief notes indicate that my health was on the decline. However, when I received my leave two weeks later, I decided to go skiing, my favourite sport. I skied for several days, which always served as a boost. Friends commented on my improved personality and colour. I spent the rest of my leave at home.

While at home my health began to deteriorate again, which led me to seek the advice of our family doctor. (I was experiencing dizziness, tension, light-headedness, persistent chest pains, shakiness, headaches and nausea.) There was also another new problem that I became aware of and that was a choking tightness in my throat, making it really awkward to get a breath of air. At times I thought that I was going to suffocate. I was constantly swallowing to try and alleviate this frightening choking sensation. My family doctor assured me again that I was physically OK and prescribed some Diazepam tablets, which at that particular time helped me a great deal.

February 1986
I returned to Aviemore and started work as a barman. Unfortunately, I don't have any diary notes for this month, so I am working largely from memory. I can remember that I had to go on another course of antibiotics because the NSU returned. I can remember the trips to Inverness with dread. It was a thirty-mile drive and my best friend at the

time, who was a New Zealander, would drive me up and drop me off at the hospital and later meet me in the local shopping centre.

On one occasion, while at the clinic in Inverness, I took a turn from which I thought my body was going to pack up and I was going to die. It was to be the first of many of these attacks. On this occasion it was so frightening that I jumped from my seat in the waiting-room and I anxiously asked the receptionist if I could see a doctor immediately as I felt so very ill. I felt as if I couldn't breathe.

In some ways, I suppose it was similar to an asthma attack. I was literally gasping for breath and I felt very weak and trembly. The receptionist made sure I was seen next. However, the doctor seemed unsympathetic and was only interested in my urethral problem. I asked if it was possible to be seen at Casualty, but he told me that was unlikely and to go and see my own general practitioner.

I remember going back into Inverness to meet my friend and still I had that choking sensation. I felt ghastly: shivery, trembly, nauseous, panicky and very, very tired. We got back to Aviemore and I went to bed — a place where I had come to spend more and more time, certainly more than could possibly be considered normal. I went back to the doctor's surgery again and this time I saw a different practitioner. He listened to my problems and prescribed some powerful anti-depressant. He also said that my problems could be resulting from the Coxsackie virus and he asked me to attend the following morning for a blood test. However, I failed to keep the appointment because I was so tired that I couldn't get out of bed. I never took the anti-depressants he prescribed.

So my illness started to run a pattern where I would generally feel nervy and tense. The tension behind my neck was particularly bad. My neck was almost stiff at times as a result.

Generally I could still manage at work, although I was not the same person I used to be. When my chest and stomach became particularly tight, I would take a Diazepam tablet (I only took one when I felt that I really needed one), but by the end of February 1986 I certainly was not feeling any better.

Another occasion I can remember is once when my brother and a friend came up to Aviemore for a skiing holiday and one Saturday evening we drank a lot of beer (which made me very sick) and this undoubtedly drained my body. I had been hoping to show the guys what a really good day's skiing was all about, but I was so ill that I ruined the day for everyone. I actually suffered a panic attack on the mountain — the first of many.

At the end of February I followed this episode on the mountain by breaking my shoulder while skiing in very poor conditions. It was probably a blessing in disguise as it meant a month's rest at my home in Paisley.

During my stay at home I decided that this would be the perfect time to try and arrange to resit my management examination. However, I remember the occasion I chose to go to the college. I had become chronically tired (something I just put down to being in a run-down condition). I was annoyed that I had let myself become so unfit. This was something I promised myself I would remedy when I returned to Aviemore. En route to the college I met a couple of friends and I could barely speak to them. Everything felt so unreal. It felt as if I was not really there, but was more like an onlooker.

When I reached the college I went to the library to meet my old classmates and again experienced this unreal feeling. I seemed to have lost some co-ordination and I felt terribly ill. After a very short time I made up an excuse to leave because the conversation was too draining for me. Although I had looked forward to meeting some of my old friends again, I now just wanted to get home to bed.

Needless to say I was unable to complete the task I had set out to do, i.e. resit the examination. The "faint" attacks came and went but generally I always felt light-headed, with a kind of brain fog, trembly and nervous. I think that is what made me convinced in the first place that the problem was "nerves".

When my broken shoulder had healed I travelled back to Aviemore, hoping that things would improve for me.

However, I remember stopping off at Pitlochry and everything seemed so unreal and I felt as if I was taking part in a nightmare.

I had certainly picked a good time to return to work as it was the Easter period and we had nearly 500 people staying at the hotel. I really tried very hard at work (in fact I was given the barman-of-the-week award twice running), but my body was giving out warning signals that I could not ignore for much longer. I seemed to do nothing apart from work, go to the doctor, or sleep. I found it very difficult to get out of bed at 11 a.m. to go to work.

I discussed with my GP the possibility of seeing a specialist at the hospital and he told me that a psychiatrist visited the village once a week and he could arrange for me to see him on his next visit. I agreed, as I was willing to try anything to make me feel better.

The psychiatrist informed me that the GP had assured him that my problems were not of a physical or organic nature and he thought that my panic attacks and other problems could well be sparked off by my fear of having contracted a venereal disease. The psychiatrist advised me to book into a psychiatric hospital in Inverness for a little while. However, I refused and decided that if I had to leave Aviemore I would return home to my parents. I really thought that everything would be sorted after I had been home for a little while. The psychiatrist in the meantime prescribed anti-depressants for me, but these never agreed with me.

I moved back to my parents' as soon as I could, but my health did not improve one bit in the month that followed. I was unable to resit my examination at college because of ill-health. All I seemed to do was sleep and suffer from bizarre symptoms which doctors persistently attributed to anxiety.

I decided to ask for a consultation with a psychiatrist to try and get this so-called "anxiety" problem sorted out once and for all. Fortunately I was appointed to a young specialist who was very understanding.

His advice was that I take regular physical exercise. After "trying" to play a game of five-a-side football, for example, I could hardly make my way home. It was explained to me that the lactic acid in the muscles now built up a lot quicker than it used to, but this sounded slightly too technical to me. When I played tennis I would usually win the first set and then start to lose. Then I would be ill for a few days and have to stay in bed. It was in the summer of 1986 that I became aware that my muscles would fatigue very rapidly and I was finding it difficult to do things that I had not even had to think about before.

The psychiatrist then started repeating blood tests every fortnight and found persistent "high titres" of the *Varicella zoster* virus. This specialist had attended several lectures on ME and became quite sure that this was what I was suffering from.

He made a hospital appointment for me in Glasgow and it wasn't till December 1986 that I was taken in for a muscle biopsy and blood tests. From the results the diagnosis was confirmed as chronic post-viral fatigue syndrome.

In his letter to my GP the specialist informed him that he had found high titres of the *Varicella zoster* virus, and that there was widespread mitochondrial damage. He advised that I supplement my diet with Efamol Marine capsules. His prognosis was that unless I recovered within two years it would be most unlikely I would ever achieve full recovery.

The psychiatrist who was the first to diagnose my illness as Myalgic Encephalomyelitis has continued to give me counselling since his diagnosis in the summer of 1986.

Early in 1987 I heard about Jan de Vries and decided to consult him. I informed him of the definite diagnosis of ME and Mr de Vries confirmed this and indicated that my ME was very active. For just over a year now I have been under his treatment and his optimism and enthusiasm has kept me sane. Although greatly improved, my health is not yet back to what it used to be.

4

Candida Albicans

THE STAFF AT the slimming department in our clinic asked me to see a very large lady who weighed well over twenty stone whom they had tried to help, unsuccessfully, in her efforts to lose weight. They genuinely believed that this lady had been most co-operative and had followed the dietary advice conscientiously, but all to no avail. They were baffled by the lack of success and therefore referred her to me.

During an interview with this lady we discussed her lifestyle and general health and it became clear that she had a *Candida albicans* parasite. The symptoms of a *Candida albicans* parasite are various, but it is a lesser-known fact that metabolic action can be severely impaired, as was the case with this lady. That the treatment of the *Candida albicans* parasite was successful soon became evident when this lady started registering steady and considerable weight loss. More important, however, was the fact that she became a much more healthy and energetic person. I still see her occasionally and without any great effort she now keeps her weight stable.

When she first heard the diagnosis she asked for further explanation as she had never heard of it before. Indeed, the *Candida albicans* parasite has never received a great deal of publicity and few people are aware that all of us carry a parasitic yeast or fungus in our bodies. On the whole this is nothing for us to worry about, unless this fungus is activated due to an inferior diet, the use of steroids, antibiotics or the contraceptive pill. Then this particular parasite can indeed become destructive.

If and when this parasitic fungus becomes active, the severity of the repercussions depends on the ability of the immune system to deal with it. If we lack immunity the fungus may flourish. It is sometimes said to be one of the mysteries of modern medicine, but I believe that because it is a dormant condition and therefore difficult to diagnose, it will develop stealthily and may get out of hand before we have been fully alerted.

A *Candida albicans* parasite can manifest itself in different ways. For example, the overweight lady had a bloated and distended abdomen, chronic vaginal thrush as well as problems in the urinary tract and bladder, causing frequent attacks of cystitis. She also experienced depression and irritability, chronic constipation and a variety of fungus-type rashes, especially on the scalp. In other cases complaints may include heartburn, abdominal pain, indigestion, flatulence or diarrhoea. There may also be inappropriate drowsiness, tiredness, swelling of the joints or heart palpitations. I have noticed that in many diagnosed cases of a *Candida albicans* parasite, people report feeling unwell after eating grains, most particularly wheat. Other reported symptoms include acne, sinusitis, nervous complaints, loss of memory, headaches and let us not overlook allergies.

The symptoms may be aggravated by tobacco smoke, perfume, hairsprays or petrol fumes. Depending on working conditions, sometimes headaches or migraines seem to be the first indications. Vaginal fungus infections or thrush and disorders in the urinary system could also be an indication that the fungus is active. However, clinical evidence is essential before a diagnosis can be reached as the symptoms are so varied that they could equally indicate another single disease or illness.

The anti-*Candida* therapy in conventional medicine will usually consist of drugs which contain Nystatin. In alternative treatment the accent is placed on a long-term approach aimed at keeping this situation under control. Therefore the treatment may take a little longer but, especially by reviewing the dietary habits of a *Candida* patient, we feel we have a better chance of controlling the problem and avoiding its recurrence.

I mentioned earlier that it is not unusual for a Myalgic Encephalomyelitis patient to also suffer from a *Candida albicans* parasite and vice versa. This combination always makes the treatment more intricate and delays recovery. The same goes for the ME patient who has previously suffered from glandular fever.

On a recent trip to the United States and Canada I lectured to thousands of people and had endless discussions with colleagues in both alternative and conventional medicine. It was amazing to learn how many people have suffered from a *Candida albicans* parasite. In fact it is now widely recognised in medical circles that antibiotics themselves could well trigger the *Candida* infection. I will come back to this possible connection in Chapter 8.

Before treatment can be effective the dietary management and possible drug use of a *Candida albicans* patient must be looked into and also the environment where he or she lives should be investigated. Areas of mould or mildew in the home or at work must be located and eliminated. These may be found in the kitchen, in food cupboards, left-overs and the refrigerator. In living areas check rugs, damp cupboards, pot plants and mouldy books. In bathrooms and toilets, carpets and areas under the sink etc. should be checked.

There are a number of factors that could be relevant for the *Candida albicans* sufferer. It is important that the condition of a *Candida albicans* patient is regularly monitored by a doctor or practitioner and that the patient co-operates fully to overcome or control this problem, because if the *Candida* infection is allowed to continue unchecked a chronic deterioration can take place in the general health of that person. We often find that people are prepared to co-operate quite well up to a certain point, but when it comes to a change in diet, they are invariably reluctant to change their habits. I sometimes think that it must be easier to change people's minds on their political affiliations, their religion, or even on their choice of marriage partner, than to change their daily diet.

Nevertheless we must realise that the majority of *Candida albicans* patients are suffering from an allergy which can develop into a very serious problem. Therefore the following factors have to be taken into consideration for treatment to be successful:

—repairing the intestinal tissue;
—re-establishing a better intestinal bacterial flora;
—strengthening the immune system;
—eliminating the fungus by anti-fungal agents.

Dr Alfred Vogel never tires of stressing the importance of the friendly bacteria which are frequently destroyed by harmful bacteria. Lactobacillus will take care of the friendly bacteria threatened by chemicals, food additives, antibiotics and other destructive factors. We are fortunate in that a variety of remedies are available for re-establishing an optimal intestinal micro-flora, which is so important to the *Candida albicans* patient. This fungus, after all, resides mainly in the intestines although it chooses to establish itself in the vaginal regions and on the skin.

Sometimes I have been able to make an early diagnosis of *Candida albicans* when throat and mouth problems have been reported. A saliva test is a safe and reliable method to check if a *Candida* infection is present.

Patients often ask what they are supposed to do when *Candida albicans* has been diagnosed. Their main consideration is how they can quickly get rid of the irksome rash. When given the options, they begin to get a little anxious because they mostly want a quick recovery without having to instigate major changes in their dietary management. Sometimes I am happier when I recognise a strong aversion or inability to change. I would rather know from the beginning what their attitude is, than be given false promises.

It will already be of great help if they succeed in cutting out yeast products, sugar, wine, mushrooms and cheese. But there is more we can do. If all fermented products were to be banished, even quicker results will be achieved. This category contains alcohol (including wine), all kinds of cheese, smoked fish and meats, sausages, hamburgers, hot-dogs, nicotine and milk. People tend to get slightly confused and include yoghurt in this category because it is a dairy product and fermented, but natural yoghurt is actually very beneficial, especially when it is live yoghurt containing the acidophilus bacillus, which is used in yoghurt culture.

In cases of a severe thrush I have had to act contrary to my principles as a naturopath and warn the patient against the use of fresh vegetables and fruit. Although at any other time I am very much in favour of these, it is essential that they are avoided until the thrush has cleared up. At that point patients are again more than welcome to use these excellent products.

Using the known anti-*Candida* supplements and the acidophilus powder, commonly found in health shops, we can encourage the friendly bacteria to stimulate the digestive system into further activity.

I doubt if it is general knowledge, but I can inform you that the micro-organism in the average bowel weighs anything from 3 to 5 lbs — a not inconsiderable weight. The more we can stimulate the friendly bacteria present there, the better we can keep a fungus problem under control. A tremendous anti-*Candida* agent is one of the oldest-known herbs, namely garlic. Among other remedies suitable for the treatment of *Candida albicans* infections, garlic as well as olive oil has proved to be of great help. For serious problems I do not hesitate to prescribe castor-oil treatment also.

From the excellent book *Oils and Fats*, written by my sometimes co-lecturer Udo Eramus, we learn how important it is that we use the right fats. I attended one of his lectures recently in Canada, called "Fats that heal and fats that kill". He made it very clear that even flaxseed oil can help restore the balance in the essential fatty acids. Hence my compulsion to extol the benefits of the Oil of Evening Primrose, especially for cases such as those we are dealing with here.

The Nature's Best company has some marvellous products in their range to help overcome the *Candida albicans* problem, as well as to enhance the function of the immune system in general. The newly developed product Probion, from the Lamberts range, is often prescribed by practitioners to ward against side-effects of the extensive use and abuse of anti-microbial drugs. Probion helps to stimulate intestinal peristalsis, which in turn helps to physically remove any pathogenic micro-organisms, toxic bacterial metabolytes and waste products. Probion also increases general resistance to infections and has anti-carcinogenic properties.

Another excellent new product in the Nature's Best range, which is often prescribed for especially allergic conditions, is Imuno Strength. This formulation of important nutrients gives an added boost to the immune system.

The daily use of some Molkosan — a milk-whey product made by Dr Vogel and available from Bioforce — in salads or general cooking is also advisable. This, too, is a formidable anti-*Candida* agent.

Whenever the immune system is under threat I would recommend taking the following supplementary vitamins daily:

—one to two grams of vitamin C;
—B-complex (yeast free);
—zinc (50 to 100 mg);
—magnesium (300 to 400 mg);
—vitamin E (200 to 400 IU);
—selenium (50 mg);
—folic acid (30 mg);
—vitamin A (10,000 to 15,000 IU).
(In cases of a disturbed sleeping pattern one gram of Tryptofan may also be taken.)

In addition to these very helpful supplements let us recap the essentials as mentioned at the beginning of this chapter:

—Avoid all sugar and refined grain products, especially wheat.
—Avoid all fermented food and drinks.
—Avoid all yeast products.
—Follow a low-carbohydrate diet, but make sure that it has a high fibre content.
—Avoid alcohol and nicotine.

Especially if constipation occurs, as is often the case, it may be advisable to use an enema. Although I do not advocate the excessive use of enemas, in cases of chronic constipation I do not hesitate to recommend any of the types detailed below. However, it is important that the instructions are adhered to.

Remember that the over-use of enemas may result in weakening of the bowel tone. They should be used only as part of a complete purification programme and only for as long as is absolutely necessary. An enema can also prove invaluable whenever signs of impending colds, flu or digestive problems appear.

General instructions for herbal enemas

Make a strong infusion of herbal teas or a decoction of roots and bark. Strain and cool. Use two teaspoons of these herbs per pint of water. This infusion may be made up in advance, but should preferably be used within twenty-four hours, though certain herbs will keep for up to seventy-two hours. However, once souring or scum appears, throw it away. The infusion should be stored in a glass container in the refrigerator or other cool place. Make herbal infusions or decoctions in stainless steel pots only.

Properties of herbal enemas

Catnip enema

Mildly calming, soothing and relaxing.
Effectively brings down fevers.
Excellent for use with children.

Camomile enema
Excellent for recuperative periods after an illness or a healing crisis.

Detoxifying enema
Make a decoction of yellow dock and burdock roots, then add red clover and red raspberry infusions.
This stimulates the liver to dispose of bile, thereby relieving stress and pain in a healing crisis.

Slippery-elm enema
Mucilagenous, soothing, softening and nourishing.
Excellent if the patient is having trouble eating or retaining food, as the bowel absorbs the nutriment.

Sage enema
Warming and purifying.

Garlic enema
Liquidise four cloves of garlic in one pint of warm water and strain.
Profoundly purifying and an excellent aid in the treatment of worms.

Astringent enema
Use witch hazel, bayberry or white-oak bark.
Used to help stop diarrhoea and dysentry.

Flaxseed enema
Relieves inflammation, pain and bleeding.
Even more effective if two teaspoons of liquid chlorophyl are added.
Also aids the general healing process.

If the kidneys and/or liver are affected due to a *Candida albicans* infestation, Harpagophytum or Boldocynara from Dr Vogel may be used. Either product can be extremely helpful to patients, especially when they seem to have reached a crisis point.

Of the many patients with *Candida albicans* I have treated some have been known to also suffer from Myalgic Encephalomyelitis. One particular patient was willing to copy some notes from her diaries for a case history and has given her permission to have them included in this book. It concerns an intelligent and sympathetic 29-year-old unmarried lady and from her notes you will see that the quality of her life was gradually being eroded because of the *Candida albicans* infestation.

Let me tell you straightaway, however, that she has since fully recovered and it is an absolute treat to see her so happy and fulfilled. In her treatment the accent was largely placed on stimulating and restoring her immune system, which had obviously suffered for a considerable length of time.

Extracts from the diaries of a Candida albicans *sufferer*

Age 15 years
Without any obvious reason suffered from sudden bouts of tiredness and lethargy. Swollen neck glands.
Doctor prescribed iron tablets.
Gradual improvement over several months.

Age 21 years
After donating blood was ill for two full days.
Then slept for 24 hours without being able to wake myself up.

Age 22 years
Tracheitis.
Lethargy, tiredness, weight loss, nausea. Pins and needles. Weakness in limbs (mainly on right side). Disturbed sleeping pattern with vivid dreams. Feeling "flu-like". Mouth abscess.
Visual disturbances — blurred vision. Eye test showed marked deterioration.
Swollen lymph nodes in the neck.
Sudden episodes of tachycardia for no reason.
Cold extremities, poor circulation and lower back pains.
"Strange" feeling.
Poor concentration and impaired memory.
Depression, which was completely out of character.
Regular menstruation pattern — no amenorrhoea.
Feeling helpless and desperate.
No one can find anything organically wrong.
Diagnosis was "anxiety state", although there was absolutely no reason for anxiety. I enjoyed my work, had a good social life and home life and had no worries out of the ordinary.

I took prescribed antibiotics, but refused any other medication, e.g. Valium, Propanalot, Opilon and anti-depressants. Unable to work for six months.

Hospital tests included:
test for glandular fever — negative;
routine blood test — normal;
barium meal and follow through — normal;
ECG — normal;
all X-rays and other tests — clear.

Conclusion from medical consultant:
Thought the problem may have been caused initially by a viral infection.

After six months felt able to cope with work, but never felt 100 per cent, or as well as before.
Tried to dismiss it altogether. Didn't like to think or talk about it.
Still felt tired. Developed boils on the back periodically, whereas I had never had skin problems before.

Experienced periodically:
puffy eyes and face;
being very pale and tired and suffered remarks from friends to this extent);
vague abdominal pains.

Age 25 years
Tired and lethargic.
"Strange" feeling again, with sickly headaches.
Blood tests taken by the doctor showed no anaemia, but he nevertheless prescribed iron tablets.
Made to feel once again like a neurotic.

Age 27 years
Right foot, especially big toe, badly swollen. Red and hot and very painful, although no injury. Diagnosed by doctor as clinical gout.
Blood test and plasma within normal range.
Felt generally unwell.
Sudden hot flushes.

Three months or so later
Weakness in arms and feeling "strange" again.
Very tired and lethargic.

Internal tremors and visual disturbances.
Fear of something terrible going to happen. Worse when out in crowds and mainly occurring during the day. Always feel better at night.
Sickly headaches again.
Neck stiffness and giddiness.
Painful right shoulder and right side of back.

When at college I occasionally felt agitated and fidgety in class. Couldn't concentrate and wanted to walk out.

Age 28 years
Sudden development of allergy.
Symptoms clearly noticeable on wrists and abdomen. Also worse under both arms, round the mouth and on left eyelid.

Recently
Since the end of last year, feeling extremely tired, trance-like — sort of going around in a daze.
Finding it difficult to cope at work and to get up early.
Finished course at the end of January and have been unemployed since.
Tiredness and lethargy continue.
Apathy.
Extremely "fed up". Feeling sorry for myself.
Feel as if I am never going to improve.
Afraid of doing anything or going anywhere, in case I feel ill or something worse happens.
Terrible fear of having a "fit" or something.
Frightened of "losing control".

Luckily, after following the advice she was given and taking the medication, this patient enjoys excellent health, as I mentioned earlier.

5

The Immune System

EVERY SO OFTEN we read in the press stories relating to allergy syndromes and new allergic reactions that have been discovered. We have read about people having to spend their lives in oxygen tents because they have become allergic to breathing in our everyday air. Admittedly this is one of the more extreme examples, but most certainly we hear so much more nowadays about asthma, hayfever, allergic reactions to plastics . . . I could go on and fill a page with influences to which people have shown allergic reactions. This is before I have even started on food. Why does it seem that we have gradually become so much more vulnerable?

It is not just allergies that we read and hear about. There are articles on "acid rain" and its frightening effects on our environment. We read about gases that damage the ozone layer, due to which the average temperature on earth will increase in the future. We read about damage caused to marine biology because of pollution of the water. We read about radiation accidents and the dangers of nuclear waste disposal. We read about germ warfare and related tests that make the whole affected area uninhabitable for decades.

Why, then, are we surprised if healthwise we are reaping the fruits of the seeds sown by man? We cannot blame creation; we can only blame the way we choose to live our lives. We pump chemicals into the air with our exhaust fumes from cars and power plants, thus disturbing the atmospherical balance. We chop down whole rainforests, disturbing the ecological balance. We really cannot blame anyone but ourselves. We try and make life easier by indulging in convenience foods, overlooking the fact that these are mostly full of chemicals in our delight at having to spend less time in the kitchen. We do not protest when the manufacturer adds artificial colouring to our food and our drinks, because it looks more attractive.

Why, then, are we so surprised that in the long term we have to foot the bill? Did we really think that we had grown so clever that we could get away with it?

In my opinion it is a relatively simple matter. Because of these environmental, ecological and mostly man-made influences our immune system cannot cope. We are subjecting our immune system to a very gradual process of erosion, without realising that we are in collusion. It is like a conspiracy. Let us get on with it now and enjoy the short-term benefits and let the next generations worry about the long-term after-effects! In effect, we have not wanted to consider the repercussions and have closed our minds to them.

Unfortunately, though, it is we ourselves who are having to pay the price. Our immune system is not capable of resisting these outside influences and that is the reason that we read frequently about AIDS, *Candida albicans,* the post-viral fatigue syndrome and similar illnesses.

We have an obligation to stimulate and care for our immune system and we have the collective responsibility to keep this earth a place worth living in.

Even a frequently occurring influenza or cold can be a signal that there is something not quite in order with our immune system. Can this really be due to the way we live and what we eat? After my many years in practice I really do believe so.

I will try and explain in simple terms the function of the immune system, which is housed in the lymph nodes and the bloodstream. The white blood cells defend and protect us from invaders. That is their responsibility. The T-lymphocytes and the B-lymphocytes have to work together with the immune defence proteins called antibodies. The body's defence will immediately suffer if the T-lymphocytes or T-cells are impaired and perhaps lose their ability to function correctly, as they have to be constantly alert to be effective. Any time that something is out of balance in our body, possibly only a simple cold or another minor health problem, the immune system is called upon. It is at this point that we could encounter problems and the reason for this is quite obvious.

The immune-strengthening nutrients that activate our immune system are not always readily supplied. It is not only the vitamins, minerals and trace elements we depend on; I have already stated that the three forms of energy are food, water and air. Well, we know that the air we breathe is polluted and our drinking water has chemicals added to it, so that leaves our food. I would have thought that this factor is very much in our own hands and that we should be able to rely on its vital energy being intact.

No exceptional mathematical talent is required here to realise that grains, vegetables, fruit and nuts contain live energy if nature has not been interfered with. From tests undertaken on produce supplied from our own organic nursery, it became clear how many more vitamins, minerals and trace elements the organically grown food contains than food produced with the aid of artificial fertilisers.

The provision of a sufficient quantity of well-balanced dietary protein plays an important part in maintaining a healthy immune system because we know that protein deficiency can lead to a lowering of its efficiency. As protein is made up of amino acids, these are important to boost the immune system to fight against all the viral invaders.

Quite correctly, then, emphasis is placed on the sufficient and regular intake of amino acids, essential fatty acids, vitamins, minerals and trace elements. Nowadays these are readily available in a wide range of supplements, but let us not overlook the importance of eating and drinking our food as naturally as possible. This way we can help the immune system to respond naturally and thus defend ourselves.

I say "natural" food because a lot of our food contains toxins through colourings, preservatives and other additives, as well as the use of pesticides and insecticides. These could cause the immune response system to turn against itself, which can lead to diseases such as nephritis, colitis, diverticulitis, anaemia, allergies, etc. A major problem nowadays is that our food is refined to such an extent that it has become deficient in live energy. It is also this live energy that is attacked by radiation, which also interferes with our immune system.

Let us just stop here a moment and consider just one example — that of wheat. This excellent product, supposedly one of the finest foods given to mankind in the Creation, has undergone so much doctoring that it now contains only one-eighth of its original live energy. As a result of these procedures wheat now contains so many chromosomes that many people have become allergic to it.

Our immune system not only helps us to combat disease but also restores our health. The white blood cells secrete substances that control viruses and fungi. Antibodies support the white blood cells, which then in turn aid the production of cells. Our immune system is controlled by hormones, secreted by the endocrine glands. This very same system is capable of preventing even the most serious diseases, like cancer, from taking hold. In my book *Cancer and Leukaemia*, I have explained how the endocrine system works and pointed out, for example, how important a role the thymus gland performs. This gland is very important for immunity, and its efficiency depends very largely on the provision of balanced nutrients.

Without adequate nutrients the essential vitamins cannot be absorbed. We often see that because of poor nutrition problems such as *Candida albicans*, deficiencies, lack of absorption and indigestion appear. These underlying problems all have the potential to turn into a major health risk.

Although vitamins do not provide energy as is often claimed, they do serve as body-builders and will act as potential prevention to disease. An old naturopath once told me that he believed germs to be his best friends. He has proved the wisdom of his convictions because even though he has reached a ripe old age, he never has a cold or flu. He is possibly one of the few remaining pioneers of naturopathy left in Britain. He started out as a young man with people like Stanley Lief, Thomson and others, and today still maintains that germs are his allies as they strengthen his immune system. He is a great believer of eating plenty of raw food and maintains that it is this which gives him his energy and resistance.

We must always remember that our health is our own responsibility and no one else's. If we contract a disease it is very possibly more self-inflicted than we would like to think, because we have neglected to care for our immune system.

In many cases we may have inflicted the damage ourselves and become allergic to something. We may have introduced tranquillisers, antibiotics and other drugs. These may be a medical necessity at times, but should never be taken indiscriminately. We will probably have neglected the fact that the natural defence system of the body, when it is balanced, will never fail to amaze us. It is the innate energy in the body that cures and not the treatment. With that positive approach and knowledge we can overcome tremendous problems, if repair is called for.

As I have said, many of today's problems are a result of the self-induced failure of the immune system. By rebuilding the body's immune system we will take the first steps towards the prevention of disease and we can only reach a permanent cure if we start from that angle.

When the body is able to produce Interferon — an isolated natural nutrient manufactured by the body in a bacteria and/or virus, breast cancer and tumours can even be controlled. The immune Interferon produced by the T-lymphocytes helps the many viruses to work with the DNA master molecule in its genetic work. The healthy cell which is fed by balanced nutrients will produce anti-viral proteins and bacteria and viruses, can defend those cells and build resistance to any illnesses and disease.

When patients are under mental or physical stress and strain the immune system has to work extra hard. This will cause the body to produce less Interferon, thereby rendering the body's immune response deficient. In such a situation it will not be possible to resist an invading virus properly and as a result we encounter problems such as the post-viral fatigue syndrome, and possibly even more serious problems in the longer term.

In the immune system the T-cells often act as a helper to the B-cells in the production of antibodies which attack invaders and protect the tissue in the body. However, when the immune system is weak, problems such as infections and toxicity can arise. It is then that degenerative diseases can appear and therefore good care must be taken to influence the immune system favourably.

Remember that we owe it to ourselves to enable the immune system, wherever possible, to deal with viruses, bacteria, parasites and toxins. Viruses are 100 to 200 times smaller than bacteria and once they have penetrated our defences and are in the body it is almost impossible to destroy them. My great-grandmother already recognised the importance of an effective immune system and she compiled a herbal formula to help the immune system. This she passed on to me and I in turn gave part of the formula to Nature's Best. They added further important components and now produce it under the name Imuno Strength. This remedy has proven its value many times over.

Another aid in the destruction of invading viruses is Dr Vogel's remedy Echinaforce, the extract of the plant echinacea and a marvellous natural antibiotic. I have witnessed many times with my own patients how successful this remedy is in getting through to the immune system. I sometimes suggest that garlic or another remedy derived from the butterbur, called Petaforce, also from Vogel's Bioforce range of herbal remedies, are used in combination with this remedy.

I have often spoken to eminent immunologists and virologists and pleaded with them to consider the use of these simple remedies. Some have been won over by the results, whereas others have been too set in their ways.

I remember an expert fisherman from the north of Scotland who was asked to advise on fishing in Ghana. While out in Africa he contracted some unknown virus which almost proved disastrous for him. Occasionally I still see him and I always consider it a small wonder that this man survived.

He was brought to our clinic by his wife and I immediately realised that he lacked any natural immunity to the virus in question. We had to fight to keep him alive. Very likely the natives were conditioned to the food and water he had shared with them and they had probably acquired an inbuilt immunity to the virus that had attacked this man. He suffered from continual diarrhoea and the virus had severely impaired the efficiency of the white blood cells in their task. I prescribed Echinaforce and my own formula of vitamins, minerals and trace elements, which is marketed under the name Probion. I am convinced that these remedies saved his life. He was fortunate in that his immune system was basically in good condition, although it had not been able to ward off this particular foreign invader.

In a report in the *Lancet* of 23 May 1987 I read that during tests on eighty-five children, nine were found to have a lack of antibodies because of drug abuse on the part of their mothers or because of infections. Inherent immunity is of tremendous importance and epidemiological data was therefore based on the predominance of the parents. Some of these children were immuno-deficient and this type of background has to be taken into account when deciding on the necessary therapy.

Everywhere in the world scientists actively search for ways to protect the immune system. Professor Karl Asai from Japan discovered the life-enhancing properties of the mineral germanium, with which I have worked for quite a few years.

The subject of germanium and its properties has been explained at length in my book *Cancer and Leukaemia*. I would feel helpless in the treatment of various diseases without being able to prescribe germanium, but I also know that it serves as a boost to the immune system.

How wonderful that nature offers us minerals such as germanium which enriches the body's own oxygen supply, destroys toxic free radicals and discharges poisons from the system that have become such formidable opponents to the health of mankind.

Organic germanium has a potent immuno-stimulatory effect and is also a powerful healer. Sometimes it is described as "a four-dimensional mineral that serves as one of the outstanding immune enhancers".

The immune modulations of germanium stimulate the T-cells to produce circulatory lymphocytes. The lymphocytes' Interferon generate macrophages. Germanium thus produces an anti-tumour effect through being a tremendous aid to the T-cells, the natural killer cells or macrophages. Therefore in cases of viral infections germanium has proved invaluable.

I have pointed out already that scientists all over the world are researching aids to the immune system and further good news has come from Japan where Dr Yukie Niwa has spent a great deal of time researching an enzyme called superoxide dismutase, often referred to as "SOD".

Like his compatriot Prof. Karl Asai, Dr Yukie Niwa realises that oxygen is essential to life in any shape or form and that the white cells need oxygen-free radicals to kill off invading bacteria and viruses in the bloodstream. "SOD" is the body's own controlling system and so, to keep the free radicals in check, it is sometimes necessary to use this. After years of hard work, Dr Niwa has produced Bioharmony, a product from natural food substances, each of which has some ability to counteract free radicals, substances also known as anti-oxidants.

The reason that Bioharmony is such a unique product is because the anti-oxidant properties which lie largely untapped in raw constituents are fully available in this finished product. Dr Niwa has trained as an immunologist and his research was funded and supported by the Japanese

Ministry of Health. It has now become generally accepted that the immune system is of tremendous importance in one's general health.

I cannot state often enough that the body has its own defence system that is given to us in order to protect the internal system against invasions of bacteria and viruses and maintain good health. It is our individual responsibility to take good care of this system.

One reason why we have been hearing so much in recent years about immune deficiencies might perhaps be the striking feature that a virus is capable of turning the body's immune system against itself; what it should be protecting becomes instead the object of its attack.

I have stressed already that we are obliged to defend ourselves and have also mentioned a number of methods which we can employ to do so. Remember, the first rule is to adhere as closely as possible to the laws of nature. Claude Bernard emphasised that "at all levels of biological organisation, survival and fitness are conditioned by the ability of the organism to resist the impact of the outside world and maintain constant within the narrow limits of the physio-chemical characteristics of its internal environment".

6

Food Additives

LATE ONE EVENING I arrived at Central Station in Glasgow on the last train from Birmingham and was waiting at the entrance to be collected for my trip home. Railway stations always fascinate me with their constant buzz and crowds of people and even at that late hour there was still some activity going on. I couldn't help noticing one poor creature who was looking round in a most disturbed manner and appeared both unhappy and uncared for. He caught my attention several times and the feeling grew that I knew him. Finally I placed him as an ex-patient of mine, although I remembered him as a handsome young man.

When he had first come to see me he told me about his problems, which had started to affect his relationship with his wife. He had approached me for help and of course I promised to do my best for him. I liked him and thought he would be co-operative, but was soon disappointed to find out that he was not really trying. Apart from some psychological problems he proved to be a typical victim of food additives, to which he had grown allergic. Unfortunately, those additives were present in his favourite foods and he was loath to forego these. I tried to reason with him and advised him on his diet and lifestyle, but to no avail. He showed no commitment because of a weakness in his character and I told him that without his co-operation there was no use wasting his money or my time any longer. He had to look elsewhere for help or suffer the consequences and he now seemed to have chosen the latter option.

With the best will in the world neither his mother, wife, nor myself could help him. His doctor had given up on him and he obviously had now chosen to live the way I witnessed late that evening at the railway station. My heart went out to him and I had just about decided to go over and speak to him when I saw him being led away by a person similar in appearance to himself. I was left with the thought: another one of the many victims of unwillingness to make some changes in the dietary pattern.

Is it any wonder that in our work we are so concerned about the chemical additives in food and that we urge people to use organic food and study the labels before they buy?

My friend Maurice Hanssen has written a very informative book, *E for Additives*, which gives excellent guidelines and information on this subject. I certainly do not want to expand on this, other than to stress how important it is that everyone be aware of the many additives in our food.

Indeed there do exist harmless additives, but equally there are additives that are detrimental to our health to such an extent as to be the cause of a breakdown in health such as happened to my ex-patient I saw that night at the railway station.

An EEC report published in 1974 listed 1,700 flavourings: 500 natural, 700 artificial, and 500 that were provisionally considered safe. Given that this figure was reached more than a decade ago I would imagine that the total number may have doubled by now. We could be fooled into believing that because these additives are listed they must be safe, but I still feel extremely wary of them and fortunately I am not alone in this scepticism.

I also read in an excellent booklet produced by the Soil Association that we in the United Kingdom consume an average of about 5 kg or 11 lb of these chemicals each year. There are some people who may even consume three times that amount and we should all be made aware of the disastrous results this could lead to.

From a recent publication by the London Food Commission, *Food Adulteration — And How to Beat It*, we learn further interesting facts. According to an extensively researched A to Z of the various ways food is adulterated, British food is the sick food of Europe. Whereas Norway has banned the use of all artificial colourings in food and the United States allows only seven, the United Kingdom permits an alarming sixteen.

Very often additives or preservatives are used to extend the life of perishable foods. If the laws laid down by the government's health authorities are not stringent enough to comply with our suspicions, pressure ought to be brought to change these laws.

In the same publication we are also warned about labels that proclaim "No artificial colours or flavours", or "All natural ingredients", as natural additives are often no better than the artificial ones they replace.

As an example of this it is brought to our attention that many manufacturers are replacing the notorious artificial colouring tartrazine with a natural colouring, annatto. Yet a Scandinavian study of patients allergic to a variety of additives indicated that more of them reacted unfavourably to annatto than to tartrazine. At the same time, however, it is pointed out that annatto is a common ingredient of margarines and as such is consumed by millions of people without ill effect.

I have already stated that the figures mentioned in the 1974 EEC report would very likely have doubled by now and in fact we find that the London Food Commission's publication lists an astonishing 3,500 additives as flavourings. As these do not have to be identified on food labels, this presents an incredible and potentially serious loophole in the laws controlling additives.

Nuclear irradiation of food is also dealt with and it is pointed out that irradiation is considered a good thing by some as it kills harmful bacteria. Other experts, however, insist that it is bad, as it will kill the micro-organisms that produce an offensive odour warning that the food has become unsafe for consumption. It may also reduce the vitamin content and possibly produce mutations of chemicals with unforeseen consequences. Irradiation of food other than for specific medical purposes is currently banned in this country, but no test has yet been developed which can establish whether or not food has been irradiated.

Among the many patients I have treated for allergic reactions to additives was a seven-year-old boy who had become very difficult and destructive, with a disturbed sleeping pattern. It was discovered that he was allergic to a yellow colouring used in orange juice and even now if he takes a sweet containing that colouring agent he goes completely haywire.

There was also a young girl who had become over-active and had suffered considerable damage to her teeth. The root of her problem was finally traced back to the colouring present in fish dressing.

It may take a long time and a painstaking process of elimination to pinpoint the offensive ingredient. How much damage is created by food additives no one knows, but daily in my clinic I am confronted by patients who clearly suffer from allergic reactions. We then start the process of elimination to see what triggered a breakdown in their health. Considering our naturopathic principles, we of course always advise patients to eat food which is as fresh and as natural as possible.

The preservation of food requires sugar or sugar products, anti-oxidants and further chemicals. Colourings are then added to make the food look more attractive and palatable. I was utterly shocked when, during a visit to a meat-processing factory, I became aware of the extent to which preservatives and colourings are used. I seriously doubt if the eventual customer ever stops to consider the possible consequences of what they are buying.

I readily admit that the preservation of food is often necessary, but it is sad to acknowledge that so much junk is deemed essential to make that food more saleable and appealing to the customer. My advice, therefore, is to read the labels and some of the information that is freely available on request from various sources in order to understand more clearly the dangers we are exposed to.

Not so long ago I treated a gentleman who ran a successful and thriving market-gardening business. He had begun to suffer from a severe skin rash and his visits to doctors and dermatological specialists had been in vain. Finally I discovered that he was allergic to a colouring additive he used in his business and which served to make his vegetables appear more attractive and green. It is sad to think that we may be purchasing such produce over the counter at the greengrocer's or supermarket, unaware of the chemicals that have been used in its production.

On one of my visits to the United States I was part of a group given a guided tour round a food-processing plant. Another member of that group was a great friend of mine and also a well-known allergist. We both realised that some of the additives used in the processing procedure there are considered to be carcinogenic.

I well remember an elderly female patient who used to attend our clinic in the Netherlands and who always made a point of buying our organically grown vegetables on her visits. It was worth watching her while she inspected the quality of the produce. One day she purchased some lettuces and the sales assistant pointed out to her that there was a slug in one of them. The old lady remarked that if a slug could live in it, she could live on it. With this remark she caused considerable hilarity, but how wise she was!

Carrageen (seaweed/Irish moss) has hardly been considered to be harmful and yet I know of a young girl who suffered from colitis and the cause was traced back to this substance. It just proves how careful we have to be. If we were aware of the possible repercussions we would do anything to avoid becoming chemical victims. Let us then also realise that excesses of sugar and salt — the most common additives of all — can be detrimental to our health. To extend the life of perishable foods, very often additives or preservatives are used, and if the laws laid down by the government's health authorities are not stringent enough to comply with our suspicions, pressure ought to be brought to change these laws.

Chemicals that are harmless in themselves can cause serious harm when they combine themselves with human serum proteins and can certainly weaken the immune system. We would be wise to investigate the possible long-term effects of, for example, synthetic chemicals, as we know that problems or even illness can result from an adulterated diet.

More and more concern is expressed about food intolerance and allergies and often much time is taken up by finding out exactly what the offensive factor may be. We are aware of some of the obvious ones, such as monosodium glutamate, which is considered to be a possible asthma agent. Certain colourings are also known to cause urticaria, migraines or double vision. I always like to double check on the doubtful cases and we are fortunate in that we have the methods for doing so in alternative medicine.

Oddly enough, I once discovered that vitamin drops were the cause of allergic problems in a young baby. This must appear ironic in the extreme, but these drops contained an additive which the baby could not tolerate. Once the cause of her problems had been established, they were overcome quickly and successfully.

A plain and simple headache may be a wonderful indication of a problem somewhere, even though the symptom is unwelcome. Regular occurrence of a headache, then, serves as an alarm signal and could well persevere and become migrainous in order to point us towards a previously unknown sensitivity.

In a similar way I found out that nitrate used in the curing of bacon was a decisive factor for a teenager patient who had started to suffer from epilepsy. Extreme allergic reactions were registered to the nitrate in foods of which she was extremely fond.

In this world of new technology I suppose there is a limited need for food additives. However, we must not let the availability and ease of convenience foods diminish our responsibility to our own health. Let us train ourselves to look at the labels and learn about what possible damage we could innocently be inflicting on ourselves.

In the next chapter, which deals with allergies, we will see that with desensitising methods and prompt action major problems may be averted and that by strengthening the immune system we can help the body to cope with the influences of additives.

7

Allergies

SOME TIME AGO a mother asked me for advice about the health of her young daughter. It was clear that the little one was in poor condition. The mother told me without hesitation that, according to her, the youngster was suffering from some form of allergy. Unfortunately, the mother had become neurotic about this and had approached an allergy clinic where samples of the little one's hair were tested. She was given a long list of possible allergies along with a sizeable bill. Not being much wiser she then took her daughter to another clinic where blood tests were taken. There she received another list of possible allergic tendencies, together with a list of recommended remedies and yet another bill. Still not satisfied, she attended a further clinic with her daughter where she received further advice along with a further invoice.

The subject of allergies appears to have become quite fashionable as a topic of conversation. Practitioners have their own opinions about it and clinics seem to be sprouting up spontaneously. Therefore, this mother could have gone on for quite a while longer, but a friend had persuaded her to approach me. She showed me the various diagnoses the previous clinics had provided her with, but when I did some simple tests of my own on the child I came to the conclusion that the little one suffered from scurvy. In this day and age this must sound incredible, but as the mother had eliminated so many foods in her efforts to pinpoint a specific allergy, the child had actually become undernourished and now suffered from malnutrition. I am fully aware that malicious intent was out of the question and the reason was more likely to be ignorance and her following irrelevant and conflicting advice.

Yet in 1988, despite our ever-increasing knowledge about nutrition, the danger of malnutrition or too restrictive and one-sided a diet, is more frequent than we would believe. I must agree that allergic tendencies do indeed appear to be increasing, but at the same time must warn that it does not pay to become obsessed about this. I know that it is possible to become allergic to nearly anything, yet I also know that we can imagine ourselves to be allergic. The other day I had a female patient asking me if she could possibly have grown allergic to her husband! It is not unknown for new allergies to be created, while at the same time we should not underestimate the problem.

Often degenerative diseases are initially caused by allergies; nevertheless I must stress that this is often due to a weakening of the immune system. If our reserves were more adequately taken care of and if we paid more attention to leading a life closer to nature, obeying its laws and investing sensibly in our health, there would be fewer allergies. What we witness today are mostly the fruits of a considerable change in lifestyle: the availability of processed foods, convenience foods and the often excessive use of preservatives are but a few contributory factors.

Various descriptive terms are commonly used when discussing allergies, such as allergic syndrome, allergic sensitivity or allergic tendency, but what really is an allergy?

The dictionary I consulted described it as an abnormal sensitivity to a specific substance. Not only is it possible to become allergic to foodstuffs, but also to atmospherical influences, pollutants or pollen for example. The last item is probably one of the most widely known factors as it causes hayfever, and I could hardly be accused of exaggeration when I say that everybody has heard of that dreaded condition.

Different methods can be used in the approach to an allergy, but if the immune system is not dealt with first the problems are often allowed to develop further. We can block allergies or treat allergic reactions with medication, use desensitising methods or homoeopathic or naturopathic remedies. It is most important, however, that we stimulate the immune system to spontaneously overcome or block the allergic reactions to whatever might be the trigger.

It is not always easy to follow an additive-free diet, because it involves a great deal of checking for possibly disagreeable additives in order that they may be eliminated. Remember, one could be allergic to a variety of foodstuffs or even non-food factors. This goes even for food substances we use daily, such as bread, which is almost impossible to obtain without additives. The water we drink may be allergenic due to ore residues or waste and here we often find the cause of such problems.

Allergists will mostly recommend that we consume more fibre, less fat and fewer processed foods and that we avoid additives, but we still have to live and, one way or another, we are bound to be subjected to allergens.

A most eminent allergist visited me for a few days and remarked at one point that I tackled the allergy problem in a totally different manner to him. I could not but agree that I probably did. He explained that he perseveres with the elimination of certain foods, but that my approach was to immediately go to work on the immune system. Much to my surprise he added that my approach appeared to be effective. Why does it work? It is not unusual to find that by rebuilding or strengthening the immune system the allergy will abate or even totally disappear. When approached from that angle many people who worry about what will happen next and what will be left for them to eat then realise that the problem may not be quite as difficult as they imagined.

It is always best to discuss with one's doctor or practitioner what the allergen could be. Avoid becoming more allergic through taking an ill-balanced preparation of vitamins, minerals and trace elements, as I sometimes see happening. This will only make matters worse. Certainly much can be done at home with the help of good advice. Perhaps the cause can be discovered by following a rotation diet or an elimination diet. However, always seek medical and knowledgeable guidance. Never treat allergies as a game or as a matter of slight importance, because a lot of misery may be prevented if approached correctly.

Just consider for a moment that under the new food regulations fifty-two additives are listed as being permissible. These additives may be contained in flour, for example, and therefore products such as bread, biscuits and cakes could very easily create an allergic reaction. Then the sometimes endless task is to discover which one of a possible fifty-two permitted additives is the culprit.

So often mothers ask me whether their child could possibly be suffering from an allergy. They continue with a list of symptoms and expect me to solve the puzzle quickly. Little do they realise that there are hundreds of possible symptoms, for example itchy skin, dizziness, unusual pallor, catarrh, vomiting, skin rashes or dark rings under the eyes. The list of symptoms is variable and seems endless. It is a matter of finding out exactly what is the matter if one feels under par. Could it be a masked allergy or possibly an absorption problem? The relevant tests will show that something somewhere in the body is out of harmony.

The other big problem of course is that we do not yet fully understand the way the immune system works. Due to changes in the immune system, a person may one day be allergic to wheat, while shortly afterwards the allergy is to milk or dust, or mite. This changing pattern is frequently experienced and therefore it is not always easy to detect an outright allergic reaction to one particular substance.

Ever since the term "allergy" was created in 1906 it has been the subject of controversy. Once a lady came to me about persistent nervous complaints. She had been to a doctor who had advised her to eliminate successively wheat, milk, eggs, yeast, tap-water, preservatives, sugar and synthetic vitamin preparations. This complicated her life considerably and unnecessarily, as it happened, as nothing was achieved by it. Like a detective I probed and enquired and finally it was deduced that this poor lady was allergic to aluminium. After she took my advice and disposed of her aluminium pots and pans and cooking utensils, she improved by leaps and bounds. In a case that was rather similar, I finally managed to pinpoint the allergen as being an emulsifier.

In yet another case the lady concerned had become almost suicidal at times. Eventually it was found that she had an allergy to lecithin, a fatty compound found in egg yolks and certain vegetables. Every now and then she would go completely haywire and it took some time before we found the common denominator. I tried to eliminate from her diet those foods containing lecithin and these amounted to quite a few. In a situation such as this the problem then arises that the diet must remain balanced to the extent that one still receives the correct proportions of vitamins, minerals and trace elements which are essential to support the immune system. If this aspect is overlooked other problems could very well be created.

Initially, this particular lady had come to seek my advice as a result of having read my book *Stress and Nervous Disorders*, in which a short chapter on allergies is included.

She followed my advice to the letter but, unfortunately, no progress was made. This was because lecithin is generally considered to be beneficial to one's health. This case proves, however, that each person must be considered individually. Certain foods can clearly act as a poison to some consumers, while the majority will benefit from their use.

Allergists are sometimes miles apart in their approach to the problem. Immunological allergists mostly direct their teaching and research towards the endogenous process and look at the steroids and anti-histamines. Other allergists will initially look at the food intake, chemicals and additives. No matter what their approach route to the patient's complaint is, the important fact is that a problem does exist which needs to be dealt with.

I am pleased to see that the chemical adulteration of our food receives much more attention nowadays. We should all be made aware of the possible repercussions of using and absorbing chemical additives and in cases of allergic reactions every detail of one's lifestyle must be studied in order to trace the offending substance.

In my work with Multiple Sclerosis patients I advocate the use of the gluten-free diet, which I compiled with the help of Prof. Roger MacDougall. Unexpectedly, we see quite often that allergic reactions disappear when Multiple Sclerosis patients follow these dietary recommendations. Many of them manage to control their conditions as a result and then of course feel much better and more hopeful.

No one knows the exact reason for allergic reactions and why one should gradually acquire an allergic tendency, but always consult a good and reliable physician who will try and trace the cause. A perfect example of how simple the solution could be was a recent patient of mine in whom I discovered a reaction to beans. It was as simple as that and sometimes we waste time by getting our priorities wrong through overlooking seemingly insignificant factors or indicators.

I have found that Dr Vogel's Harpagophytum is always of great help in allergic cases. This remedy is a herbal extract of devil's claw (*Harpagophytum procumbens*). In such cases I also regularly prescribe Urticalcin, which is a homoeopathic calcium and silicea preparation, and Kelpasan, a pure sea algae from the Pacific Ocean with all the trace elements and a natural supplement for iodine deficiency. Both the latter remedies are also available in Dr Vogel's Bioforce range.

It is wonderful to see how the above remedies strengthen the immune system and any one or a combination of these remedies can be safely used with any allergic reactions. Harpagophytum is an effective natural cleanser of the kidneys and liver and stimulates the immune system. Kelpasan supplemented by Urticalcin is well tolerated by those people who are sensitive to iodine in the kelp. Kelpasan will activate the endocrine glands and so helps the patient's immune system.

The Chernobyl accident has taken its toll and we know little as yet about the after-effects which may not yet have become apparent. This tragedy affected some people immediately and points to the danger of living near nuclear energy stations or waste-disposal plants. It is especially important for those people who fall into that category to use a combination of molasses, vitamin C, kelp and Urticalcin. That combination will always be a tremendous help as a preventative measure against possible problems that might arise in connection with the above factors. In addition, nourishing proteins and a good cleansing diet will stimulate the body to clear the toxins and eliminate allergens.

When the cause of an allergic reaction has been discovered it can then be decided which nosode, antigen or homoeopathic remedy would be most suitable. Recent tests with hayfever patients have shown that homoeopathic treatment had tremendous advantages over the usually prescribed drug treatment.

Food sensitivities often present themselves as colic, diarrhoea or in other ways, but here again the symptoms can fluctuate and be confusing. Mould or fungus sensitivity can cause a much more aggressive reaction. This can also be seen with chemicals such as petrol fumes and paint removers, as well as with cigarette smoke or gases. With dust, mite, pollens, grasses or trees there may be a quick reaction expressing itself as rhinitis, runny eyes or nose, eye irritation, eczema, sinusitis and sometimes as colon problems. Much depends on when these problems arise and whether other factors such as *Candida albicans* infection play a part.

With pesticides and insecticides instant reactions can often be noted and sometimes immediate action is called for as the results could be fatal. I do not want to be presumptuous, but I do fervently hope that somehow this book may alert all of us towards an awareness of these serious matters and hopefully stricter controls will be applied in the future.

Once I was invited to lecture together with Amelia Nathan-Hill, the author of *Against the Unsuspected Enemy*. She told me how she had battled together with her daughter and finally succeeded in eliminating the agonising and persistent illness which had threatened to ruin her daughter's life. She had suffered from depression, colitis, cystitis, thrush, eczema, arthritis, mouth ulcers, constant colds and influenza, indigestion, insomnia, swollen ankles, leg cramps, bloodshot eyes, arm and hip pains, and so on. Yet there was a happy ending to the story. It was then she decided that by writing down her experiences and trials maybe someone else would be encouraged to face up to the reality of their problem and also succeed in overcoming this onslaught on one's health. After having eliminated the offensive allergens, she ends her publication saying:

> I know we will succeed in bringing this fundamental cure within the reach of all allergic people in the struggle against the unsuspected enemy.

To overcome allergies sometimes the simple exclusion of the offending foods may be enough after they have been isolated. Yet because allergy problems are so variable other methods may be necessary and this is why I stress that this should be done only with the guidance of a knowledgeable doctor or practitioner.

For many years people have suffered from hayfever and it is wonderful that nowadays we know that with the help of certain homoeopathic remedies and dietary changes this condition can be successfully held at bay in most cases. This also goes for the condition known as allergic rhinitis. People who have suffered helplessly for years can be helped with simple natural methods. Especially in cases where the younger generation is involved, unless the situation is too drastic, my advice is to eliminate all dairy products and salt, to eat lots of honey and to take Dr Vogel's remedy Pollinosan — homoeopathic drops or tablets for sufferers of hayfever or allergic rhinitis.

Recently I was in North Africa, where this problem is very prevalent, and it was frequently remarked upon how effective the Pollinosan remedy was. Sometimes this may be combined with Kelpasan and/or Urticalcin to support the lymph glands and the endocrine glands. The patient starts taking the remedy early in the year and so builds up a certain amount of immunity and then continues to take it throughout the hayfever season.

Also here it pays to take a close look at possible food allergies, as, understandably, food can play a considerable role in one's medical history. For the immune system both the remedies Harpagophytum and Echinaforce are of tremendous value as they support and enhance the function of the T-lymphocytes.

Let us look at a product used daily and very often excessively, namely, white sugar. This highly refined product, unbeknown to most mothers, can cause hyperactivity in children. It often happens that a child is allergic to sugar and thus becomes hyperactive. In the younger child symptoms such as crying, screaming, colic and always being thirsty may be noticed. In slightly older children the symptoms may develop into an inability to concentrate, disturbing and distracting of other children, being excessively demanding of attention or sometimes being withdrawn. The start of this condition may be the point when the child becomes overactive or hyperactive.

Similarly, the above symptoms can also be observed in cases of a milk allergy or another food intolerance. Avena sativa is a fresh herb preparation of oat seeds and this remedy, together with Oil of Evening Primrose capsules, can be of tremendous help in these cases.

If the child is hyperactive due to an allergy to a specific additive, poor co-ordination, impaired speech ability or irritability may result. Frequently a shortage of magnesium, zinc, manganese and chromium is then established.

Because of a shortage of essential fatty acids and of the mineral zinc, such behavioural changes and problems can be dramatic. The child may then become ultra-sensitive and difficult to get on with.

Hyperactivity often results from additives used in the preparation of processed foods or convenience foods. Skin rashes, itching, wheezing and/or a more or less constant runny nose can mostly be attributed to any one or more of the following additives which are used frequently by the food industry: E102, E107, E110, E122, E123, E124, E127, E131, E132, E155, E210, E211, E212, E213, E214, E215, E216, E217, E218, E219, E221, E222, E223, E310, E311, E312.

With children as well as adults rapid improvement can often be obtained if a relatively simple allergic reaction is isolated, e.g. to wheat, milk, sugar, coffee, chocolate or cheese, to name but a few. So often I hear people remonstrate that they could not possibly be allergic to coffee as they have been drinking coffee for years. I then suggest that they eliminate coffee from their daily dietary pattern for a set period of time to see what happens. They often find that not only are they allergic but also addicted.

From my notes I will now give some case histories as an indication of the wide variety of substances to which one can be or become allergic. The first of these concerns a young boy who was found to be allergic to fruit juice, which caused him to break out in hives, in other words an allergic skin condition of intensely itching weals. Another side-effect he experienced was colic. Within two weeks of the elimination of fruit juice he was back to his normal self.

A baby girl was found to be allergic to bottled and packaged babyfoods, which caused her to suffer from colic and diarrhoea. Bananas also caused her to have allergic reactions. With this knowledge, these items were eliminated and the baby began to thrive.

With some patients I have managed to isolate the artificial colouring in fish dressing as being the offending factor, causing over-activity.

With an allergy to pollen the immediate symptoms are usually irritation to the lining of the mucous membranes and to the eyes.

A middle-aged gentleman who kept racing pigeons was completely cured from a developing allergy to the birds. In his case Harpagophytum and Echinaforce were prescribed and he fortunately did not need to give up his hobby of keeping pigeons, which was the joy of his life.

A lady developed an allergy to sugar which expressed itself in irksome and painful mouth ulcers. She was completely cured thanks to Molkosan, a liquid whey product produced by Bioforce Laboratories. Molkosan contains all the important minerals found in fresh whey such as magnesium, potassium and calcium in concentrated form. Molkosan is rich in natural dextrorotatory lactic acid, which in health-orientated nutrition as well as natural healing methods has a special significance. This liquid remedy offers great versatility of use as it can serve as a substitute for vinegar in salad dressings or can be taken one tablespoon at a time in a glass of water as a refreshing drink. It can also be added to vegetable juices to provide some extra zip.

A young wife and mother showed bad reactions to the contraceptive pill. This affected her walking and balance. She was successfully treated with homoeopathic nosode remedies.

Another young baby showed allergic tendencies after the mother weaned him. He became fretful, suffered from excessive colic and could or would not settle down to sleep. The mother reverted to breast feeding and he soon returned to normal. Luckily it was discovered that soya was agreeable to him and when weaned again he settled very well.

The following, by no means exhaustive, list of causes and effects also may be helpful to the reader:

- —Chocolate could produce migraines, headaches or hives.
- —Orange drinks may be responsible for diarrhoea or headaches.
- —Coca-Cola could bring on over-activity.
- —Fluoride toothpaste could be to blame for sleeplessness.
- —The colouring in sweets may cause blotchiness of the skin.
- —The chemical flavourings in crisps could produce hyperactivity.

In addition, the tremendous problems caused by a wheat allergy can range from chronic constipation to incurable skin problems. This allergy could even be a decisive factor in degenerative diseases. I am quite convinced that of the many arthritic and rheumatic problems I come across in the line of my work, quite a few could well have started by what is generally considered a relatively straightforward allergy.

Perfumes, cosmetics, petrol or exhaust fumes, paints, new carpets, wool, soap powders or detergents, wine, convenience foods, cats and dogs, are all possible sources for allergies. Let us understand that in this field we are possibly only aware of the tip of the iceberg, i.e. the more we get to know about this whole subject, the more we may have to come to accept the fact that we as yet know very little Never forget, also, that allergies can affect our physical as well as our mental health. At the same time I do not want to preach doom and sound despondent and therefore I stress that we have reason to be grateful for some of the excellent natural or homoeopathic remedies which enable us to strengthen our immune system, alongside being prepared to adapt our dietary habits.

I am reluctant to lay down hard and fast rules for a general diet in cases of food allergies. This is an impossibility as, first of all, the specific allergen needs to be isolated. To achieve this a food allergy elimination diet is very important and to this end I follow with the four phases of a diet designed by Dr McKarness, based upon his work with Action Against Allergy.

Food Allergy Elimination Diet

Phase I

For five days take only lamb, pears, Malvern water and sea salt.

Phase II

Carefully reintroduce foods as directed below. If the symptoms return or the pulse rate rises by ten beats/minute or more from the prerecorded resting pulse rate, the food introduced can be identified as the allergen. As the allergic reaction can last up to three days, return to the diet already identified as safe for three days before proceeding to test additional foods.

Day 6: Introduce broccoli and/or lettuce into the diet.
Day 7: Introduce haddock or cod and carrots into the diet.
Day 8: Introduce avocado pear and/or melon into the diet.
Day 9: Introduce tap-water and grapefruit into the diet.
Day 10: Introduce cabbage and/or celery into the diet.
Day 11: Introduce beef and/or plaice into the diet.
Day 12: Introduce apple and/or peaches into the diet.
Day 13: Introduce green grapes and herb or China tea (without milk and sugar) into the diet.

As more foods are introduced, foods which have been established as safe can be left out of the daily menu. Once a food has been established as safe, it can be eaten or left out at one's personal preference.

Phase III

Endeavour to establish a balanced diet by introducing potatoes, millet, whole rice, fresh vegetables (except tomatoes) and fresh fruit (except oranges) into the diet. Introduce each food separately on a daily basis, testing for increased pulse rate. Subsequently reintroduce pulses and rye crispbreads in small amounts and further types of sea fish than haddock, plaice and cod.

Phase IV

Test for the small amounts of foods one would wish to eat occasionally from the "Danger Food List" by checking the pulse rate after eating the test food alone against the background of a well-established diet.

Food allergies can be symptoms of mechanical or metabolic faults in the body. The elimination diet procedure will act like a therapeutic fast. Never return to a diet in which the listed danger foods are the staple food, although a healthy person should be able to enjoy them occasionally. The elimination diet procedure is more effective when preceded by an enema or colonic irrigation.

As part of the general programme it is worth remembering that physical exercise will stimulate a better flow of oxygen and therefore such pastimes as swimming, jogging or walking are recommended.

It should not be all that difficult to adopt certain changes in one's lifestyle, if only we realise that with a positive outlook positive results may be obtained.

I recently came across a perfect example of the worthwhile results of approaching a problem with a positive mind. It concerned a female patient who had been diagnosed as suffering from an irritable bowel syndrome, another seemingly fashionable disease. We come across this more and more often nowadays. The condition can be confirmed by a double-blind food challenge. This lady complained about a vague pain on her left side and it was thought to be a problem in the sigmoid area. However, the sigmoidoscopy was completely clear, though she still complained about a constant nagging pain in the stipulated area. After following the elimination diet the pain started to diminish and then we were able to establish that she was allergic to wheat and milk. We also discovered a minimal level of lead in her drinking water. With this knowledge we managed to solve her problems to her complete satisfaction.

In the past such problems have led to surgery, which rarely proved satisfactory as the causes of the initial problem had not been eliminated.

We are fortunate nowadays to be able to rely on such appliances as the Vega, the Mora or the Dermatron machines as diagnostic aids, as these masked allergic reactions can ultimately lead to depression and unnecessary misery if they remain undetected.

8

Drugs

HIPPOCRATES MAINTAINED that if all the drugs in the world were tossed into the sea the fish would suffer more than mankind. This statement seems particularly appropriate today as there seem to be more drugs and more illnesses than ever before.

In the Netherlands, my country of birth, it has been calculated that people swallow from 150 to 200 tablets or capsules a year. In the United States and United Kingdom 50-80 per cent of adults swallow a medically prescribed drug every 24 to 36 hours. One cannot help wondering if this could be reduced and if it is really necessary in the first place. There must be some doubt as to the benefits of pill-popping on such a large scale, and that is before taking into account the astronomical sums of money forked out by the National Health Service. This constitutes an unbelievable and annually increasing sum of money.

When I was still working in the Netherlands, I remember a specific project in 1974 in which research indicated that 50-60,000 people were hospitalised annually as a result of the misuse of drugs.

Of course we cannot indiscriminately discard all drugs as non-beneficial. Some drugs undoubtedly are life-savers. Many of us have good reason to be grateful for the discovery of antibiotics, but we should not underestimate the fact that antibiotics disturb the bacterial balance and some drugs cause damage and impair the functions of the liver, kidneys and other organs. The older generation among us is unlikely to have forgotten about the drug called thalidomide, which was found to cause severe malformation in unborn children.

In my practice I often see patients who are there because of maladministration of drugs. Sometimes communication problems are to blame and I would not for one minute like to be quoted as saying that drugs are not necessary. All I would ask is that a more responsible and discriminate attitude is maintained in the use of drugs. I have already said that sometimes drugs are definitely life-savers. Nevertheless, we too often come across examples of the abuse of drugs, sometimes due to errors, but more often due to unnecessary repeat prescriptions for medication. This aspect of drug abuse is frequently highlighted nowadays in the press, when we are made aware of the side-effects of long-term use of tranquillisers for example.

In *The Lancelot's Island Journal* (No. 28) it was stated that the mortality rate dropped by 17 per cent when the doctors were on strike. In Colombia the mortality rate decreased by 37 per cent during a strike, while in Israel it was actually halved. The article in the journal also stated that the medical drugs and other preparations available now number approximately 205,000, whilst new illnesses and afflictions have increased proportionally.

Relate these facts to the subject of this book, i.e. allergies and viruses and the increased occurrence of both, and then consider if there could be a connection. Often doctors are blamed and certainly pharmaceutical industries come in for their share of the blame, but whose fault is it really? We must accept the fact that our physical health is our individual responsibility and that often we ourselves could be to blame for our physical condition. It is of course much easier to pass the blame to doctors, but we cannot shirk that responsibility in all honesty.

It is the easiest thing in the world to take an aspirin or ask for an antibiotic when we have a bout of flu. But what about those countries where such medication or facilities are not available, where we see that general health is often better?

In the article in *The Lancelot's Island Journal* it was also pointed out that the present situation in the United Kingdom should serve as a warning not to follow in the footsteps of the USA. There the problems have become enormous because the drug industry is so powerful. It is therefore in a position to hinder and obstruct the progress of alternative or naturopathic medicine, whose aim it is to help people in a more natural manner. I have some very good friends of many years' standing in the United States who have only one vision: to ease human suffering in a natural way. In many cases they have been forbidden to practise and are under threat of a prison sentence. It seems incredible that this is allowed to happen in 1988.

We must guard against falling into that sort of trap here. I am certainly happy to learn that in the Netherlands there is now considerable recognition and support for alternative medicine, as is the case in other European countries, and I sincerely hope that we will follow suit.

As a trained pharmacist I am only too aware of the pressure on the part of drug companies for an increased turnover. They cannot be considered to ultimately have our best interests at heart. They are first and foremost a commercial venture and therefore it is up to us to accept our personal responsibility and be realistic about what we are doing to ourselves and our environment and what we are going to hand over to the following generations.

We may stop pain with the aid of a pill, but we will not create health with the use of a drug. We need body-builders and not body-breakers and therefore in Chapter 5 on the immune system I have concentrated on how we can strengthen and rebuild the immune system in order to stave off the attacks we are subjected to.

When healthy, the body is able to ward off invaders, poisons, toxins and viruses. This also applies largely to allergies if we care for our immune system and do not kill off what is good in the process of killing off offensive bacteria.

One of the four principles of homoeopathy composed by Dr Samuel Hahnemann, the founder of homoeopathy, states clearly:

> It is of major importance to restore harmony possibly without the use of drugs that may work in the short term, but which often destroy friendly bacteria in the process, and to investigate the natural defence system to the outside influences in order to build up the immune system and to minimise the effects of outside influences detrimental to the patient's general health.

In other words there is a vital energy or a vital force within man which can be influenced either positively or negatively. In this book I am trying to explain how important that vital force is. It is ultimately dependent on our own input, as well as being sensitive to vibrations. It can work for us or against us and we must be prepared to face up to the responsibility which we have with respect to our own health, inasmuch as that is in our own hands.

We cannot allow ourselves to be bombarded by pills, powders or capsules in order to get rid of some ache or pain. Nor for that matter should we choose sleeping tablets to get a decent night's sleep. When these methods have to be used regularly, we must look for the cause of the problem or else these drugs will be allowed to rule our lives and without these artificial means we would find it difficult to survive.

On recent visits to Canada and the United States I was surprised to discover how often Multiple Sclerosis patients suffered at the same time from a *Candida* infection. This particular condition badly impairs the digestive system and this is especially serious for Multiple Sclerosis patients as they then suffer from poor absorption and lack the necessary enzymes of which they have such need.

I then remembered a female patient in our residential clinic in the Netherlands who suffered poor health. Not only was she a Multiple Sclerosis patient, she also had a *Candida albicans,* virus which had severely affected her digestive system. On top of or possibly because of her problems, she had become dependent on a particular sleeping tablet which has since been taken off the market. These gave her suicidal tendencies. Indeed, not long after the discontinuation of the sleeping tablets a television programme highlighted the problems and produced evidence about the side-effect of feeling suicidal. Certainly, it seemed, my patient was not alone in that aspect.

The combined problems had certainly taken their toll on her health. Immediately it was decided that she should be weaned off those sleeping tablets and we could soon see a change for the better. Even though she was not too good on her legs, I remember taking her for a stroll on a Sunday afternoon a few weeks later. Her general health improved considerably after a sensible diet had been introduced and her Multiple Sclerosis condition stabilised. With the help of some natural remedies to restore her health, she left the clinic after five weeks, feeling and acting like a different person to the one we had admitted.

Although I fully admit that sometimes drugs are necessary, so often they are used indiscriminately, which had been the case with this lady. She underwent a total transformation after adopting a more natural regime and cutting out things which proved offensive to her general health. She was a well-educated and intelligent lady and it puzzled her why her digestive system had improved so remarkably. She considered the possibility that she might have become more prone or less resistant to certain germs as a result of suffering from Multiple Sclerosis. I reminded her of Louis Pasteur's belief that often disease was caused by germs and/or viruses and that keeping this in mind it was quite possible that her Multiple Sclerosis condition was initially caused by a virus. We talked a lot on this subject and I explained that personally I would not be surprised if the massive doses of antibiotics she had been prescribed for her

Multiple Sclerosis condition before this had been diagnosed as such had not caused unnecessary grief and suffering in the long run.

We discussed the meaning of the word "antibiotic", as *bios* means *life*, and therefore antibiotic means anti-life, and it is after all administered to remove infections.

Quite understandably, the medical breakthrough when antibiotics were discovered in the 1930s was deemed a miracle. Yet so often antibiotics are administered in such massive dosages that I often wonder how much damage is done in the process of healing or treating a specific and isolated condition or infection.

We can look at what antibiotics have achieved with the greatest respect and gratitude but equally we must realise that some of the beneficial bacteria are destroyed in that same process.

A very dear friend of mine was a relation to Sir Alexander Fleming and we used to talk with great admiration about the tremendous efforts he undertook in his work which led to the discovery of antibiotics. Once over dinner the conversation centred on medicine in general and I asked her opinion of what Sir Alexander Fleming, if he were still alive, would think about his discovery now. It seems to be a universal trend to indiscriminately prescribe antibiotics with the underlying thought that even if it does not help at least no harm will come of it. My friend's spontaneous answer was that Sir Alexander Fleming would most likely be horrified.

Let us consider how the created imbalances could be reversed by adopting a more positive and constructive philosophy — that of *pro-life* or *pro-biotics* — and supporting the beneficial bacteria which are an integral part of our bodies.

My friend Monica Bryant, founder of the International Institute of Symbiotic Studies, is a lecturer in nutrition at Sussex University. Whenever we have lectured together, I have listened with great interest to her explanation of the concept of pro-biotics on different health subjects. She maintains that it is essential that pro-biotics are administered immediately after antibiotic treatment. I have touched upon this already when I mentioned the Probion remedy. Antibiotics have a detrimental effect on the microbial population in the colon and often the effects of over-use or long-term use of antibiotics become apparent as depression, fatigue or diarrhoea.

The intestinal flora is divided into two main categories: the resident or indigenous flora and the transient flora, which passes through the system with food. This is enriched by the indigenous population of bifido bacteria in the colon to stimulate their growth with transient bacteria.

Lactobacillus bulgarius and *Streptococcus thermophylis* continue to foster the growth of the bifido bacteria. These bacteria produce organic acids in a number of ways, which create a protective acidic environment in the intestines, acting like a natural antibiotic and preventing the development of pathogenic infections.

Problems such as *Candida albicans,* where the colonisation of opportunist fungi penetrate the intestinal wall, so allowing micro-organisms and toxins from the colon into the blood, can result in a great variety of diseases, creating allergic conditions.

Probion, a product of the *Lactobacillus bulgarius* and *Streptococcus thermophylis,* is of the greatest help in combating this process. Very often, by combining it with Echinaforce, Molkosan and Harpagophytum, we manage to achieve excellent results with this treatment method. It means new life where other life has been destroyed and we only need to look at the skin of our patients when working with the intestinal system. The oral ecology, the skin ecology, the intestinal ecology and the vaginal ecology are dependent on the metabolic flora in the intestines and its results in the liver. It has become apparent that 80-90 per cent of AIDS patients suffer from *Candida* infections, as do a large percentage of Herpes patients. *Candida albicans* lives in everybody and in itself presents no danger. However, food and drugs such as antibiotics or steroids could stir the *Candida* into action.

The body is grateful for every little bit of encouragement. In the world of traditional herbal medicine it is well known that the blackcurrant (*Ribes nigrum*) stimulates the immune system and *Candida* patients mostly react well to this treatment. It is unfortunate that this has been known for generations and yet no account has been taken of this fact in modern medicine.

It saddens me to see people dependent on drugs, who have destroyed their health with sleeping tablets and/or tranquillisers. An investigation published recently in the *Daily Record* detailed some of the disastrous effects of this habit on the general health. Often the liver objects and skin irritations and allergic reactions become apparent. In particular the kidneys and liver suffer undue stress.

Then of course there are the so-called recreational drugs, i.e. cigarettes, alcohol or addictive pleasure drugs. Especially with drug abuse it soon becomes clear how much damage has been done to the intestinal flora, requiring additional medical attention and an effort to try and turn the tide and to rebuild and strengthen the immune system.

The withdrawal symptoms when a recreational-type drug is eliminated can be symptomised by severe pains in the stomach and bowels and these can be alleviated by a remedy from the Vogel range of herbal preparations — Centaurium. Its name is derived from the Latin name for cornflower, which is the main ingredient. It may sound a slight exaggeration, but I would suspect that there is quite an element of truth in the saying that if all recreational drugs were eliminated we might have nearly empty hospitals. I often remind patients that without a doubt nicotine is a poison that lowers the efficacy of the immune system and will very likely cause us harm sooner or later.

Alcohol, when over-used, can also be a toxin and we need only see what happens in the end to an alcoholic. It is not only the liver which cannot cope with alcohol over-use, other organs are also affected, and that is before we consider the risk of brain damage.

It is not only the "hard" drugs but also "soft" drugs that can do physical and psychological harm. It cannot be right that one has to depend on a drug for life; this can only be justified in specific and isolated cases.

Often in smokers and drinkers I recognise a craving for certain drugs. It may only be for a relatively innocent substance like aspirin, but just as easily this could be for a more harmful substance. Let us not forget either the other unhealthy addictions that exist excessively today, namely the exorbitant use of either sugar or salt.

In my book *Stress and Nervous Disorders,* I referred to a pattern of fashionable habits in medicine over the centuries. In the Middle Ages there was a thriving market for powdered residues that were supposed to improve one's health. Since that time we have had periods of burning irons, blood-letting and enemas. There was also the period when it was fashionable to remove the large intestine, supposedly to ease constipation problems, except that these then turned instead into diarrhoea problems. Even today constipation is still a major problem. Early in the twentieth century it was fashionable to remove the tonsils. Then we entered the age of antibiotics, which was followed by the age of cortisone. Possibly, at present, we are witnessing the approaching end of the era of tranquillisers and sleeping tablets.

If we had become happier and healthier it would not be so bad, but the fashion in medicine tends to come up with newer and often more dangerous therapies and remedies and at great expense we drift yet further away from nature.

Undoubtedly, like antibiotics, cortisone has done a lot of good, but I wonder if that can be balanced against its detrimental effects. Cortisone breaks up and destroys the potassium in our body and I frequently see the effects of its long-term use with patients who come to our clinic. In many cases I cannot even help them with massage or osteopathic manipulation, because the softening of their bones caused by extended use of cortisone makes that impossible.

Many of these fashionable therapies have undermined our immune system to such an extent that at a later date we will have to pay a heavy price. Many discoveries have saved millions of lives, but if we are sensible we will realise that any excess will cause problems.

I could not begin to describe the allergic reactions and the decreased efficiency of the immune system I have witnessed over the years in patients who were using the contraceptive pill. Again, this is an innovation for which many have paid the price.

I remember one female patient in her late teens where no agreement could be reached on a diagnosis and in desperation she was finally diagnosed under the all-inclusive umbrella of Multiple Sclerosis. I discovered in the end that she was allergic to the contraceptive pill. If one suspects allergic tendencies in this direction, one would be wise to stop taking it as soon as possible as in her case the results were dramatic. I remember that the doctor in charge of this particular girl phoned me to compliment me on my diagnosis. After homoeopathic treatment this girl was finally able to walk normally again.

Then, however, her doctor again contacted me to inform me that he was worried about overpopulation of the world if a campaign against the birth-pill could gather strength. Again, the responsibility lies with the individual, as it concerns our own body and we must individually ask ourselves the question as to how we can best go about looking after it.

This brings us back to the question: are such remedies and therapies body-breakers or body-builders?

I knew one doctor who was barely able to write a prescription any more as the result of an anti-inflammatory drug he had used which has since been withdrawn from circulation. He was in very poor condition and his liver was severely affected. However, with dietary adaptations and the use of natural remedies, he is now able to function properly again.

The choice is ours: do we use drugs in order to treat the symptoms, or do we find the hidden cause and treat that in a natural manner? Whether we place our bodies under undue strain, or whether we are sensible in our choice of diet and include body-builders — it is our personal freedom of choice to make that decision. If, however, we allow our bodies to regenerate, we may expect to live a more enjoyable and trouble-free existence.

9

Mental Health and Nutrition

QUITE A FEW years ago I read with great interest an excellent book written by Dr George Watson: *Nutrition and Your Mind — Eat Your Way to Better Mental and Physical Health*. The reader may wonder why I refer to such a publication in a book on viruses and allergies. In the first instance I might agree, if it wasn't for the fact that I know from years of experience how much nutrition can affect one's mental health. With that I mean possible allergic reactions to food itself or food additives and preservatives that can cause physical as well as mental imbalances, leading to problems that are hard to identify or classify.

Important studies have taken place on the effects of dietary deficiencies that have led to allergic reactions. Imbalances in food or drink can lead to unexpected situations, as Dr George Watson points out in his book. He writes about a professor who sat watching television in the evening with his wife. She got up very calmly out of her chair and without a word made her way to the front door, opened it and walked out into the road. Only by braking sharply did a passing motorist avoid running her down and the sound of the brakes alerted her husband. Shaken, he brought her back into the house and consulted Dr Watson.

To cut a long story short, the investigation brought to light that half an hour previously this lady had sipped a brandy mixed with water and sugar. Nothing untoward had been noticed and she seemed to be perfectly all right. However, alcohol is quickly absorbed by the stomach and raises the blood sugar by acting on the liver to release more glucose. The brain and the nervous system are totally dependent on normal functioning and there can be a loss of normally acceptable emotional control if insufficient glucose is available. The medical report on this lady showed that her blood sugar had gone below the minimum level necessary and the oxygen rate, which is the index for the speed with which tissues break down food to create energy, was far below normal.

The combined readings of low blood sugar and a slow rate of energy production were then connected to a low-carbohydrate slimming diet which she had been following. Due to these combined circumstances the normal function of the nervous system was not being supported. Her diet was corrected and no further incidents took place and her problems completely disappeared.

I remember reading about this case with the greatest of interest so many years ago and I have very often referred back to this book. It reminds me of a personal experience which occurred shortly after the end of the Second World War. During the latter years of the war we had all been deprived of sugar of any kind. We tried to return to a normal life and slowly wheat products such as biscuits and cake became available again.

At a certain period my school reports became so shocking that my parents became very worried and asked for an interview with the headmaster. He told them my work was still suffering, but also that my behaviour was unruly and disruptive. It was most likely that the marks on my next report would actually be worse still, as he had not noticed any improvement, rather the opposite. This of course not only disappointed my parents but also worried them greatly as they had noticed a change in my character at home as well.

In those days there was not much awareness of allergies and I was referred by our doctor to a school psychologist and finally ended up seeing a psychiatrist. He found nothing wrong with me, but informed my parents that although I was not hyperactive there were certain problems which he encouraged them to investigate further. Fortunately there were no drugs prescribed. My parents battled on in the meantime and they finally discovered the key. Because I had been deprived of sugar and sugar products for so long, I had become almost addicted to sweets and chocolates and was indeed making up for lost time. I was not given these at home but obtained them from other sources.

My parents had a very serious talk with me and tried to explain what I was doing to myself and my sugar intake was drastically reduced.

The book written by Dr Watson was therefore of the greatest interest to me and from then on I tried to collect as much information and knowledge on connections between nutrition and mental health and possible resulting allergic reactions. To my mind there is no doubt that nutritional imbalances are related to allergic reactions. I had to find out the hard way and over the years I have learned from many patients' misfortunes and have felt sympathetic to their problems through my own experiences.

The efficiency of the body's ability to transform food into energy depends on the availability of certain factors, one of which is oxidation. Young people sometimes drop out from school or university and often these could be regarded as victims of their upbringing or circumstances. They may have been brought up with incorrect dietary advice or, as they have left home and are busy with their studies, they may have fallen into the habit of surviving on convenience foods. Their addiction to certain foods can result in biochemical reactions in the nervous system or other tissues in the body and are dependent on a slow or fast oxidation rate, according to one's individuality. In this respect some knowledge of the metabolic or digestive process is advisable as imbalances in the combining of foods could then be avoided.

A correct diagnosis and investigation of the metabolic system of each individual should be carefully considered in order to find out where imbalances are prevalent or problems have occurred. The human body must be nature's finest invention and alarm bells will ring when something is not right. In my case certain sugar products will immediately result in a headache which shows that the damage done in earlier years has caused allergic reactions which I probably never will be able to shake off.

However, a measure of common sense is all that is basically needed. If we notice a change in our emotions and reactions after eating certain foods we need to eliminate these and, by doing so, find out which specific substance is offensive to our system. Medical guidance by a qualified allergist or practitioner can be rewarded by improved mental and physical health.

I have already used the term common sense and must stress here that allergies can be responsible for uncharacteristic emotional and sometimes imaginary reactions. Allergic reactions depend on positive or negative mental influences. Some allergies have found their base in one's own vitality, aggression, attitude and even in how much love we have for others compared to ourselves. This may well be the reason that if we compare two asthma sufferers with the same origin to their problems, we see that one can overcome an attack much more quickly than the other.

The allergic patient therefore has to set himself some questions as to what could be the cause of the allergy.

—"Why have I changed from being a loving and considerate person to a much more aggressive one?"
—"Could there be a connection between an imbalanced food intake and an imbalance between mind and body?"
—"Have I really changed my outlook on life and if so, why?"
—"How much do I want to be helped?"

By asking these questions we may perhaps come across the answer as to why allergies are constantly on the increase.

Are they all to blame on additives and preservatives and adulterated food patterns? Could changed morals and the acceptance of a more permissive society have any bearing?

There is a lesson to be learned from the infant that is breast-fed, when a comparatively odourless, soft and non-irritating stool is produced. Rarely do we come across constipation with breast-fed babies. If we look again at that same baby when it progresses onto pasteurised milk, we often see that it will produce more irritating and certainly much more odorous stools. Constipation will also be much more likely and sometimes infantile eczema or asthmatic reactions. When allergic reactions occur, often goat's milk or soya milk will help to overcome these problems. The initial allergic reactions may not necessarily be permanent, but they can be interpreted as an aversion to change.

With a slight philosophical touch we could consider life as a chain of desires. Even the life of a newborn baby begins as a desire. All beginning is conceived from love and desire. Unfortunately, even in young lives, the desire for inappropriate foods or allergenic foods is often greater than for the foods that are really good for us. Near enough all children, given the choice of a packet of crisps, an ice-cream, a fizzy drink, a bar of chocolate, or a square meal, will choose any but the last option. Irrespective of their illness or disease, I very often ask patients about their food preferences. So often their list contains allergenic foods or foods that cause an imbalance.

Unfortunately, media advertising is so effective that we are sometimes swayed against our better judgement towards purchasing convenience foods, often rich in additives, colourings and preservatives. With that aspect in mind, I feel justified in warning against physical and mental reactions, where mind and body lose their joint balance.

Without doubt nature teaches us to eat fresh and natural foods, if for no other reason than that these are supplied in that condition. Nature does not supply us with frozen, processed or adulterated junk foods, and it is human interference that brings these onto the market. If we allow our innate intelligence to follow its course, it will serve us as an excellent guide as to what we really should eat and drink. Often, however, it is a case of not wanting to be bothered and in our haste we quickly grab a bite without paying attention to the ingredients.

Although ample evidence is available to show a significant relationship between a physical/mental imbalance and specific nutrients, there has not really been enough research undertaken in the possible role diet plays in the early stages of disease or illness.

In a previous chapter I have already raised the point that our immunity can suffer because of imbalances. A deficiency of one single essential nutrient can make all the difference to one's physical health and one's behavioural attitude. The list of possibilities is endless and because of such an imbalance diseases brought on by germal or viral or allergic influences, due to a faulty metabolism, could almost be classified as degenerative diseases. This is a very real possibility because the attack on the immune response can be severe. A faulty body chemistry could easily develop due to incorrect nutrition characterised by additives, colouring or flavourings. Excess sugar, salt or fats could also be contributory factors. Despite risking accusations of repeating myself unnecessarily, I will once again stress the importance of keeping our food intake as natural as possible and taking positive action by eliminating allergenic foodstuffs. A positive attitude will encourage and enable the immune system to overcome many allergies.

It never fails to surprise patients who come to the clinic with a list of allergens to which they react unfavourably, that once they adopt dietary changes most of these allergies seem to disappear like snow before the sun. They will be advised to change their diet from processed, convenience and junk food in favour of food that has some life in it. It has frequently been possible to dispense with supplementary vitamins, minerals and trace elements once a balanced food pattern was adhered to.

Especially when young children suddenly become disruptive or difficult to manage, it is quite often due to an excessive sugar intake. They could easily become hyperactive and a change in their behaviour can be affected in less than a fortnight when sugar-containing items are eliminated. We must never forget that food can be a friend as well as a foe.

My friend and co-lecturer Barbara Reed gives us much food for thought in her excellent book *Food, Teens and Behaviour*. In her function as chief probation officer she looked for a possible link between food intake and the behaviour of her prisoners. She came to the conclusion that a lot of these prisoners had been food junkies in their youth. Many were addicted to sugar, wheat products and convenience foods and later turned into heavy drinkers.

She raises an interesting point in her book, stating that the area of the brain responsible for thought, learning and moral and social behaviour, starts to shut down and the brain diverts its dwindling energy resources to the brain stem. This controls the more primitive responses, i.e. the drives for food, sex, aggression, defensive instincts and basic bodily functions.

It is interesting to learn that working with prisoners and young people alike, she came to the realisation that as a result of their previous lifestyle many had become hypoglycaemic without being aware of it.

Hypoglycaemia indicates low blood-sugar in contrast to diabetes, which is characterised by an excess of sugar in the blood. Hypoglycaemia can cause mental problems, aggressiveness, nervousness, depression, forgetfulness, poor concentration, irritability, nightmares, nervous anxiety, suicidal tendencies, etc. The delicate balance here can be easily upset by an extra ice-cream, even a spoonful of sugar and sometimes with erratic children I have tested it is incredible how even a small amount of sugar can bring a drastic change in the situation.

Recently I was fortunate in being able to contribute towards saving a marriage that had always been considered a good and reliable relationship. Things had gradually gone from bad to worse due to irrational behaviour by one of the partners. It finally became apparent that this person had developed quite severe hypoglycaemic tendencies. This became apparent because every time sugar or a sweet was taken, irritability, moodiness and aggression were experienced. In this way the two partners had become totally incompatible.

I prescribed zinc, vitamin C, chromium and some other natural and herbal products. After the corner was turned and their relationship started to pick up again, they admitted that they had actually reached the stage of physically hitting out at each other. This emotion was one which they had considered totally alien to themselves and of which they were dreadfully ashamed.

This story certainly indicates how a single food substance can change a person's character if that substance is disagreeable to their body.

It is never too late to deal with a low blood-sugar problem and its balance. How does it come about? Normally, the fuel which is used by the brain cells and which we call glucose or blood sugar will work properly when the brain cells get a steady supply of blood sugar. The body normally manufactures glucose out of carbohydrates from the food we eat, the two kinds we know as sugars and starches. The body can use its normal supply of carbohydrates and from the blood sugar it makes glycogen. This is an energy product that is safely kept in the liver and the muscle tissues.

If glycogen runs out, it can even convert protein and fat into glucose. The body has enough safeguards to provide its own supplies. If, however, there is not enough glucose, the blood-sugar level depletes to meet the needs of the brain and then we get the problem of hypoglycaemia.

Logically, one would then wonder why not just take some sugar. The problem is that white sugar, or a better name for it is sucrose, is a highly refined carbohydrate. In the refining process white sugar has lost all nutrients such as vitamins, minerals, enzymes and fibre and is therefore nearly 100 per cent pure sucrose. It then enters the body very quickly and the body needs to do almost nothing to break it into glucose. It enters the bloodstream and soon reaches the intestines and is of little benefit.

Sugar found in fruit and vegetables, however, is combined with fibre and not only will help the digestive system, but has a much better absorption. Under those conditions white sugar makes the body jumpy and encourages it to respond by producing more insulin for the pancreas. The balance is then lost and often a diabetic or hypoglycaemic shock can be expected.

Often it is claimed that incorrect diets can drive people to drink and an alcohol-sugar dietary connection is perhaps the reason that many prison cells are unnecessarily occupied. If only this balance could have been corrected in time!

An endocrinologist is therefore sometimes puzzled by the instant changes in someone's character because of this imbalance. I explained the system of endocrinal glands in my book *Cancer and Leukaemia*. The pancreas is particularly sensitive to changes in food and disharmony with the endocrine glands results. If one gland is in trouble, they all seem to react. The endocrine glands are subject to physical and mental changes, while they are responsible for a balanced hormonal pattern in the human body. The system is so closely interlinked that even endocrinologists cannot always understand why problems, e.g. in the pancreas, should affect the thyroid, the thymus, the adrenals or any of the other endocrine glands.

In this role, neuro-transmitters, which can either make us happy and calm or emotionally upset for that matter, are totally dependent on the correct nutrients. Choline or tryptophan, for example, are essential nutrients, as are amino acids, and these will balance the emotions, nerves, sleeping pattern or behaviour. Very often we see with violence, repression or obsessions, that a tremendous change can take place with the use of tryptophan. It is also possible to influence one's natural tryptophan production by taking vitamin B_6 or niacin. All the amino acids or neuro-transmitters are necessary to help correct possible imbalances. The composition of foods therefore needs investigation in order to see that adequate vitamins, minerals, amino acids and essential fatty acids are available.

Allergic reactions to cow's milk or wheat can be symptomised by hypoglycaemic reactions. A glandular disfunction can be the result, which is often made worse by drug intake. It is a fact that stress and worry will influence the endocrine glands and I maintain that in many cases hypoglycaemia or diabetes results from emotional upsets or from lack of rest, relaxation or meditation. These seven of the smallest glands in the body can be responsible for a lot of unnecessary problems when their inter-relationship is unbalanced.

The adrenal glands, located just above the kidneys and responsible for the production of adrenaline, are triggered by the signals from the hypothalamus. An allergic reaction to an allergy-producing food can cause a rush of adrenaline which might cause an over-stimulation leading to a noticeable change in behaviour, possibly expressing itself in uncharacteristic violence.

A case comes to mind of a boy who was not only allergic to sugar but I also found that he reacted badly to aluminium, lead, and to the mercury in the amalgam of his dental fillings. This allergic combination made the son of settled and balanced parents into an uncontrollable, aggressive teenager. Toxic material, even toxins in our drinking water, can cause numerous problems and combined with chemicals from processed foods, can bring one to the point of no return. A balanced food pattern and the introduction of some vitamins, minerals and trace elements can fortunately change the condition of such a case quite quickly once it has been recognised.

The rising crime figures all over the world could well be attributed to an imbalanced food pattern, which leads us back to a possible connection between nutrition, crime and delinquency.

Dr Lendon Smith, author of the book *Feed Your Kids Right*, is a co-lecturer with Barbara Reed. They have presented numerous case histories to show that this process can begin from early infancy. According to them the seeds of a life of serious crime can be laid in the early years through the excessive use of sugar products. This can then later be aggravated by environmental exposures.

The criminologist Alexander Schauss is fully convinced that the present system of criminal justice is going nowhere.

Instead of spending money to seek effective rehabilitation for criminals and prevention of crime, the money is being spent on larger and more secure institutions and jails.

It just leaves us to wonder what could be achieved if we seriously researched the dietary habits of these people. I would not be surprised if we found a lack of essential nutrients required by the brain, thus affecting one's attitude, behaviour and outlook on life.

As the dietary approach has so often been shown to be defective in such cases, I am quite sure that it could play a big role in criminality. This same nutrition-based approach has proved extremely helpful with hyperactivity in children, with cases of schizophrenia, alcoholism and neurosis.

I have been surprised and, even more so, encouraged by the results of the tests Prof. Roger MacDougall and I undertook with Multiple Sclerosis patients. In my book *Multiple Sclerosis* I have explained how, for instance, the gluten-free diet helped us to obtain unexpected successes with schizophrenic patients. This again indicates the importance of our food-intake pattern to our brain cells, affecting psychological changes, whether it be a matter of improvement or deterioration. The gluten-free diet is extremely beneficial here, as the sugar intake is extremely limited and with criminal and violent tendencies this diet has also proved helpful.

Without doubt we need to re-educate ourselves and we must become more alert as to how we react to certain foods. The responsibility we have to our own body and health needs to be accepted in full. Rather than being encouraged or influenced by attractive advertisements, we must exercise some logical thinking and acquire some basic knowledge on nutrition. Barbara Reed stated in one of her reports that she had not seen one single person back in court who had followed her advice on a nutritional diet.

An eighteen-year-old drug addict with a criminal record was described as uncontrollable to me by his parents. They had given up all hope of a possible rehabilitation for him and had reached the point of despair. During tests I discovered his craving for foods that contained considerable quantities of artificial flavourings and colourings and a marked deficiency of the vitamins E, C, B_1, B_5 and B_{12}, and the minerals biotin, selenium, zinc and iron. We set out to rectify this through dietary changes and a prescription for vitamin supplements. To help him with withdrawal symptoms we recommended the herbal remedy Centaurium and also introduced Imuno Strength. Before long a change could be detected where previously no hope had seemed to exist. He now leads a contented life and holds a responsible position.

I often detect among my patients who are receiving treatment for alcoholism very low levels of vitamin C. Even then, on high doses of vitamin C alone, we can often see a change. In my book *Stress and Nervous Disorders* more relevant information is given under the chapter on Alcoholism.

People are generally puzzled as to how a simple allergy can change a person's character. Why does one become an alcoholic? Why does a person commit a crime? Possibly no straightforward answer is relevant here, because the longer I work with such patients, the more I realise that an allergy can lead to further problems. An alcoholic not only inflicts physical damage on his or her liver, but also on the brain. Certainly a crime is not committed because a person is born with criminal tendencies or because these have developed as a result of his or her upbringing. Do not rule out the possibility that the mind of the person in question might have been transformed by a food allergy, deficiencies, pollution, sensitivities, the influence of television, drugs, lack of relaxation, disturbances in the endocrine system or other self-created situations.

There is a great concept to be studied in this respect and the small glands of the endocrine system that are responsible for physical growth also carry the responsibility for mental or spiritual growth. Unfortunately these are often influenced in a negative way through ignorance.

I have interviewed prisoners sentenced to life imprisonment and one particular case struck me as being especially pathetic. I was absolutely horrified by what I learned about this person's endocrine system from an iridology test.

I have applied iridology tests to thousands of pairs of eyes, but I had never previously come across such an alarming disturbance in the endocrine system, which undoubtedly had played a major part in his behaviour. After lengthy discussions and explanations this enthusiastic man decided to adopt a vegan lifestyle, i.e. a vegetarian diet, but eating no animal produce whatsoever — no fish, eggs, cheese or milk. He has opted to channel all his efforts into regaining a balance for his "multi-split personality", as he now refers to it. He is very eager for help and therefore I intend to support him to the best of my abilities.

From our interview it became evident that many contributory factors could be found in his upbringing and environment. His mother was a gentle and kind-hearted person, without any authority, however, and involuntarily major errors were made in his younger and formative years.

Although the phraseology very likely did not exist in those days, he was definitely hyperactive. When I asked him how his mother reprimanded him when he had been naughty he told me that after a ticking-off she would stroke his head and give him some sweets or a bar of chocolate. Little was she to know that this would make him more hyperactive, allergic and occasionally aggressive. We went through all the allergy tests in the presence of either the prison governor or officers and they were astounded when, after the tests, I pointed out to them the substances to which he was allergic. The three major ones that became apparent immediately were sugar, wheat and milk.

It was interesting that he was not allergic to alcohol and he admitted that he was not very fond of alcohol. He was, however, very keen on the three above-mentioned substances and admitted to being addicted to those. I had already noticed that he would sweeten both his coffee and tea with several heaped spoonfuls of sugar and admitted to eating up to half a loaf of bread at a time. His allergy to milk became obvious when I learned he was asthmatic and used inhalers to keep this condition under control. His allergic reactions manifested themselves as hypoglycaemia and his greatest wish was to overcome his dark and depressive moods.

During the various interview sessions we had his overwhelming desire to be able to live a normal life was clearly evident. He hated being a criminal and was encouraged when I informed him that to my mind no one is born bad, but that criminal tendencies are mostly due to influential circumstances.

The endocrine system is very much subject to emotions, although endocrinologists do not always admit to this. In this prisoner I clearly recognised a great desire to be helped. It was striking that when he was given some highly concentrated white sugar during tests the whites of his eyes became noticeably whiter.

In previous publications I have already pointed out that there are seven endocrine glands and I often wonder if this number is significant, as the colours in the solar spectrum also amount to seven, as do the light receptors in the retina of the eye. The eyes are great mirrors of what happens, for instance, in the pancreas, the adrenals, the thyroid or any of the other glands and all these will display any lack of harmony.

When investigating the background and habits of prisoners who serve long, possibly even life sentences, we often find a similarity in their dietary habits during their formative years. To me this is again an indication as to how we should approach the problem of over-populated prison cells. I am equally convinced that a similar study would show that many hospital beds are almost unnecessarily occupied.

Intensive scientific research into the field of nutrition and its action and reaction on the mind is more than justified according to the knowledge we have gained so far. This would open up enormous possibilities in a field where, as yet, we have only touched the tip of the iceberg. We have at present only an inkling as to how instrumental proper nutrition may be in relation to people's mental health and condition.

With great regularity new publications appear on improved dietary habits and much knowledge can be obtained through reading these. Let us, however, first and foremost understand that we should aim to keep our food as natural as possible and use nature's bountiful supplies to create energy. If we intend to follow the laws of nature then we should immediately scrap junk food, additives, preservatives, colourings, alcohol and nicotine.

There is a lot of truth in the saying that the simpler the meals are kept, the more beneficial they are. The proper combination of foods is most important and if this is done correctly, it will undoubtedly lead to a better digestive and metabolic process. To this end I list below some of the more important rules to achieve a good balance:

—Avoid eating carbohydrates along with acid foods.
—Never combine vegetables and fruit.
—Avoid eating concentrated proteins with concentrated carbohydrates. It is better to eat these during two separate meals.
—Do not consume two concentrated proteins at the same meal. This means that two different proteins such as nuts and cheese should not be combined. It is better to eat them during two separate meals.
—Do not consume fats with proteins. Again it is better to split these over two separate meals.
—Do not combine sweet foods with proteins, starches or acids. It is again better to separate these, because one has to be careful to avoid fermentation, which is inevitable if sugars of any kind are delayed in the stomach by digestion of starch, protein or acid food.
—Do not use too much fat.
—One concentrated starch meal daily is sufficient.
—Acid foods may be eaten with sub-acid foods.
—Sub-acid foods may be eaten with sweet foods.
—Salads may be combined with proteins and starches.
—Chew your food very well.
—Drink after, but not during the meal.
—If one still feels hungry after a meal, this very likely is an indication that the balance was not correct.
—Bottled or spring water is healthier and better than tap-water.
—If possible, choose organically grown foods.

—Use foods which are as wholesome as possible.
—Wash vegetables gently and thoroughly.
—Try not to mix hot and cold foods and drinks.

On the same subject more information can be obtained from the following books: *Nature — Your Guide to Healthy Living,* by Dr A. Vogel; *Food Combining,* by Doris Grant and Jean Joyce; and *Food, Teens and Behaviour* by Barbara Reed. Obviously there are many more worthwhile publications available on the importance of food combining and proper dietary management, advising us on allergies and the elimination of foods. Those listed above only serve as examples as I have found them to be very informative.

To close this chapter I will quote a statement made by a psychiatrist: "No illness that can be treated by diet, should be treated by other means."

10

Conclusion

HARPAGOPHYTUM IS KNOWN in our language under the name of devil's claw. We sometimes wonder how descriptive any given name is, but in this case it is extremely apt in that devil's claw really seems to put its claws into the problem, i.e. it reaches out to the heart of it.

Harpagophytum has been mentioned several times already in earlier chapters as a remedy for different problems that have been dealt with. In treatment programmes for viruses, parasites and allergic reactions, the results achieved by using Harpagophytum have often been beyond my expectations. I have also described this remedy in my book *Traditional Home and Herbal Remedies*, as for centuries African natives have used this root for its medicinal properties. It was not until this century that it became available for wider use.

The reason for the name devil's claw is that this particular plant displays finger-like growths that are protected by thorns and are of unusual appearance. The medicinal attraction is that it contains no toxic substances. As a result it is suitable for all-round cleansing of the organs and it has a good diuretic influence. Because of its cleansing properties it acts as a preventive measure against possible undesirable conditions.

This herbal product has no side-effects and Dr Vogel is of the opinion that devil's claw is the most frequently used medicinal herb worldwide. It is in fact true that we generally prescribe Harpagophytum in combination with one or more other remedies. For the conditions dealt with in this book it is the most frequently prescribed product as it takes care of metabolic dysfunctions and as such boosts the immune system. For the same purpose I often combine this remedy with Imuno Strength, a product from Nature's Best, to further strengthen the immune system.

In some African countries the Harpagophytum root is used even for insect bites. I can remember my mother warning me that flies, for all that they look so harmless, are likely carriers of disease and infection. Insects, of course, apart from possibly spreading infection, certainly have a more instantly painful sting and in this respect the native tribes in Africa could teach us a thing or two. Where once insects were regarded as carriers of diseases and therefore a major threat to our health, this has been overtaken by bigger problems such as allergies and pollution.

Allergies can and often do result from polluted food substances and therefore we must decide whether specific foods are hostile or friendly to our health. Our intake of food and drink is decided by ourselves and thus there is nothing to be lost by becoming label-conscious so that we may come to realise what is agreeable and what is disagreeable. This is what we can do ourselves and then hope that governments will take up the gauntlet and lay down stricter guidelines for the use of chemicals. Scientists and nutritionists will then have the responsibility to stop a further deterioration in the quality of our food due to pesticides, insecticides, chemical additives and pollution.

A few days ago I was interviewed for an article in a well-known international magazine. The reporter asked permission for a personal question. He wanted to know which country, given the choice, I considered safest environmentally and where I would choose to raise my children. I was sad to have to tell him that I did not think such unspoilt countries exist any more. Considering the three forms of life energy — food, water and air — we realise that perhaps food can be controlled and possibly so can water, but not the air we breathe. Switzerland used to be considered a "safe" country, but even there trees in the forests are dying as a result of pollution. As that is a fact, what chance is there for us?

The rubbish we burn disappears into the atmosphere. We know about acid rain and the Chernobyl cloud will affect us worldwide. It is sad to say that there is no safe place left and what we therefore must do is arm ourselves against these constant attacks by strengthening our immune systems.

An ex-employee of the atomic plant Windscale who, of course, is bound to silence by the Official Secrets Act, has nevertheless spoken about his concern, as he considered the risks of pollution to be irresponsibly high. Only by combined efforts can we hope to control the situation. If we can unite universally to minimise the pollution problem, we may be able to ward off the further deterioration of nature as we inherited it. Personally I travel and lecture throughout the world to advise, help and to fight the battle, so that every effort can be made to protect health and environmental conditions.

It was with delight that I read in one of the Sunday papers an article written by Peter Spinks reporting on a project to safeguard the quality of drinking water. The article was entitled "Clean-up in the field of drinking" and reads as follows:

> Researchers have developed a method for stopping fertilisers polluting water supplies in farming areas.
>
> In Britain, a billion cubic metres of ground water need treatment for nitrate pollution every year, and water authorities only have rudimentary denitrification plants. These use materials which cause contamination in their own right.
>
> The new system — developed by scientists at the Agricultural University of Wageningen in the Netherlands — uses a column of resinous material to extract nitrate particles,

produced by fertilisers, and replace them with harmless bicarbonates.

The nitrate is then consumed in another column by special bacteria, turning it into innocuous gas. The water is then recycled.

As the bacteria do not reach the water there is no danger of contamination. The result is waste-free water. "It tastes rather like spa water," says project engineer Abraham Klapwijk. "Lots of people quite like it."

The system, for which a patent is pending, is essentially self-sustaining and re-usable.

The British Water Research Institute in Stevenage is awaiting the results from a £75,000 Dutch pilot plant in Doetinchem which purifies 14 cubic metres an hour. The system has already proved cheaper than finding alternative drinking supplies and would cost approximately two pence more per cubic metre if applied in Britain.

Of course it is wonderful that we are trying to develop methods to avoid fertilisers polluting the water, but why should it be necessary? Surely there are effective natural methods that would save us from becoming a drug or chemically controlled society?

Homoeopathy and herbalism offer us a choice of remedies suitable to control influenza or a cold. Therefore measures such as those reported in *The Observer* of 26 June 1988 would seem not in the interests of the public's general health. The health correspondent, Annabel Ferriman, writes:

Drug firms "guilty" on flu scares:

Three flu vaccine manufacturers have been found guilty of breaking the pharmaceutical industry's rules regarding advertising to the public. They have promised not to do it again.

The Association of the British Pharmaceutical Industry has reproved Evans Medical, Duphar Laboratories and Servier Laboratories for promoting flu vaccine to the public at a press conference last October.

The manufacturers, who banded together to set up the Influenza Monitoring and Information Bureau, held a press conference at which they said that Britain was likely to face a serious flu epidemic in the winter. Subsequently, some newspapers carried such banner headlines as: "Nine million at risk of killer flu".

Flu vaccine rapidly ran out, leaving many people in risk categories — old people, in particular — unable to get vaccinated.

In the event, the number of flu cases reported last winter was extremely low — only 500, against 1,000 in the winter of last year, 2,300 for the two previous years and 5,000 for the epidemic year of 1975-6.

The Association of the British Pharmaceutical Industry lays down that British drug companies should limit press conferences on new drugs or vaccines to medical or similar specialist publications.

The Association's code of practice committee upheld a complaint by Social Audit, a pressure group which monitors the drugs industry, finding a breach of three clauses of the code: banning statements designed to encourage the public to ask their doctor to prescribe a product; banning putting out information about medical products through the lay Press; and restricting invitations to press conferences to the medical, pharmaceutical and scientific Press.

The industry's association said the manufacturers had assured them that similar breaches of the code would not occur in the future.

So many people nowadays have become dependent on tranquillisers or sleeping tablets and we appear to have turned into a society of pill-poppers. Other addictions freely indulged in are alcohol, nicotine, soft or hard drugs. These drive us ever further away from nature.

In our clinic I see people who in their innocence have been poisoning themselves over a number of years and I can only say: "I hope they are forgiven, because they did not realise what they were doing."

Let me assure you, however, that nature itself is our very best ally and healer and even though we may have entered a situation where no relief seems possible, a number of simple and effective natural methods are available to reverse the process. Some willpower and common sense is all that is required, along with medical guidance.

I look at the trees and I think with a sad heart that even now scientists are still divided on the subject of how pollutants are the cause of losing whole forests. Let us take steps, so that some of our beautiful nature — trees, plants and all else that grows in nature — may be saved. All these things were created by God for our existence and survival. Have we been good guardians? It is said that first the forests will die and then the people. Well, then, we must make haste and stop the process of no return.

Plankton is one of the best suppliers of oxygen and yet it is rotting away in our oceans. Despite that, we still continue to adulterate our drinking water by adding chemicals.

May I remind you of the aim of the World Health Organisation: Health For All by the year 2000. We will not achieve this if we keep eating sweets and chocolates and drinking fizzy drinks made to look appetising by adding artificial colourings. No goodness whatsoever is obtained from such products.

Somewhere I read that the public health service of the sixty-seven poorest developing countries, e.g. China, spend less on their total health care than the rich Western countries spend on tranquillisers alone. In view of such facts, how can we possibly set our aims so high as to achieve Health for All by the year 2000?

The Director of the World Health Organisation advised us "to rise to our own health responsibilities, by eating wisely, drinking in moderation, no smoking, driving carefully, taking sufficient exercise, learning to live under the stress of life in general and by helping one another to do so".

The advice our forebears received when there was something wrong was: water, air and sunlight. Today, unfortunately, this very old and valued naturopathic principle has lost most of its benefits. These wonderful gifts of life have been interfered with by mankind to such an extent that so many allergies and viruses have encroached on people's lives.

Thankfully, at the same time nature has supplied us with some wonderful remedies to minimise the effects of what we have brought upon ourselves. I have already pointed out one prime example of this: Pollinosan, for the hayfever sufferer.

A short while ago I received a letter from a grateful female patient expressing her thanks for helping her during a severe attack of shingles. With the help of homoeopathic remedies and plant enzymes she managed to overcome her problem within three weeks.

Dr Vogel, in his book *Nature — Your Guide to Health Living*, lists seven enemies of good health:

1. fungi
2. bacteria
3. viruses
4. parasites
5. diseases caused by deficiencies and allergies
6. over-eating
7. poisons

He expresses his regrets that, despite the advances in science and technology, mankind is not well enough informed to be able to recognise and evaluate the numerous enemies to human health. I do hope, however, that we recognise our obligation to protect ourselves from harmful influences. In this nature can assist us, as it has supplied food for our existence and herbs for healing.

I was pleased to learn from an article in the *Glasgow Herald* of 5 August 1988, that concern among doctors as well as consumers has led the British Medical Association to set up a working party to investigate residues in food. The article, written by Jennifer Cunningham, reads as follows:

> Although some work on the effects of pesticides and fungicides has been done from the point of view of health and safety at work, very little has been done in relation to the health of ordinary consumers. We know what happens when people have short-term, high-level doses of substances like paraquat, but we do not know what happens as a result of long-term, low-level exposure. There is good evidence from wildlife of the need for proper investigation, for example, from the number of birds of prey killed by pesticides at the end of the food chain after feeding on small animals

who have eaten affected crops. One study in America has already shown that 60 per cent of herbicides, 30 per cent of insecticides, and 90 per cent of fungicides in common use induce tumours.

A report by a research committee of the Government's Crop Protection Commission due to be published this month, and leaked to Friends of the Earth this week, makes clear that the levels of pesticide and fungicide residues in fruit, vegetables and grains are likely to be much higher than thought and that there is no reliable, systematic and scientific testing of stored crops. Friends of the Earth have added their criticism of the UK regulations on maximum residue levels (MRLs). Despite a new set of regulations for fruit and vegetables to be instituted by the Ministry of agriculture from the end of the year, significant items, such as lettuce, celery and potatoes, have been left out in the levels for certain substances. For example, no MRL for dithiocarbamates — a group of fungicides — has been specified for lettuce and celery, although the United States Environmental Protection Agency has described one of the group, approved in the UK for use on lettuce and celery, as "probably a human carcinogen".

The Ministry of Agriculture's line yesterday was that the limits for cereals and products of animal origin which came into effect on July 29th are required to be in line with EEC limits, but there is no such requirement for MRLs on fruit and vegetables, although the Government has decided to impose new limits anyway from December 31st. . . .

Even the familiar potato, stored happily for generations, is now doused with a sprouting-inhibitor and fungicide by the name of Tecnazene to a greater extent than in other countries. The substance is another which is missing from the Ministry of Agriculture's list, with no reason given but almost certainly to postpone taking a decision which will go against the growers. A WHO working party recommended a maximum level equivalent to 1 milligram per kilogram, but the level effective for sprout control in British storage conditions means some potatoes will come out at levels above that.

Already Marks and Spencer is rejecting potatoes with residues above that level and the other image-conscious retailers are likely to follow suit. The whole issue is a big one for the increasingly vociferous consumer lobby with the combination of health and food being a particularly powerful vehicle for a campaign.

Friends of the Earth lost no time in making their point: "These disclosures strengthen our case for the deindustrialisation of food production and for financial incentives to get farmers off the chemical treadmill," said Andrew Lees, FOE's pesticides campaigner, but the need for more research and an obvious and central government body to take responsibility for monitoring and control of residue levels and advising on regulations and practice is an obvious one, mentioned by each of the report's sub-groups. That is the very least that ought to happen.

Although I see many cases of wheat allergy, it would be unfair to claim that wheat in itself is not an excellent food substance. Wheat was given to us by God in his Creation, but unfortunately with the use of artificial fertilisers the wheat grain as we know it is now a mutation containing a multitude of chromosomes, giving rise to allergic reactions.

On my recent visit to North America I had the privilege to meet some Amish and Mennonites who are members of an evangelical Christian sect and whose aim is to live as closely to nature as possible. Their whole agricultural system rests on organically grown produce and wheat allergies do not occur in these communities because their food substances are pure and unadulterated.

On 8 July 1988 the headline of the *Toronto Star* read: "Heat and smog has choked Metro". The article informed us that the previous day the temperature had soared to a record reading of 37.6° C, or 99.6° F. The air pollution index soared way past the "acceptable" 32 mark and beyond the health hazard level of 50, to peak at readings such as 74, 84 and 86 in various districts.

High temperatures can be responsible for an increase in pollution index figures, but even with low temperatures the pollution index figure is too high.

At the beginning of this book I wrote about the farming couple whose health had been affected due to pollution. I had an opportunity to speak with them earlier this week and with admiration I listened to their determined protestations not to let the matter rest. They do not fear being taken to court in their fight for human rights. I hope they can encourage others to follow their conscience and take a stand.

During my last lecture trip to the United States I was invited to speak at a health conference in Las Vegas. After I had spoken quite poignantly on these subjects, I was handed a little note by a charming young lady. The note contained only a few words, but with these I would like to close this chapter as well as the book:

There is nothing common . . . about common sense.

Bibliography

A. Vogel *Nature — Your Guide to Healthy Living,* Verlag A. Vogel, Teufen, Switzerland.

George Watson — *Nutrition and Your Mind,* Souvenir Press Ltd., London W1N 8HP.

E. Cheraskin and W. M. Ringsdorf — *Predictive Medicine,* Keats Publishing, Connecticut, USA.

Theron G. Randolph and Ralph W. Moss — *Allergies — Your Hidden Enemy,* Turnstone Press, Wellingborough, England.

Barbara Reed — *Food, Teens and Behaviour,* Contemporary Books,Chicago 11, USA.

Ross Horne — *The Health Revolution,* Ross Horne, Avalon Beach, NSW, Australia.

Leon Chaitow — *Candida albicans,* Thorson Publishers, Wellingborough, England.

Robert Eagle — *Eating and Allergies,* Futura Publications, Camberwell, London SE5.

Richard MacKarness — *The Food Allergy Plan,* Unwin Paperbacks, Boston and Sydney.

Günther Schwab — *Bij De Duivel te Gast,* De Driehoek, Amsterdam, The Netherlands.

John Hamaker and Donald A. Weaver — *The Survival of Civilisation,* Hamakers Weaver Publications, Michigan, USA.

Paul van Dijk — *Gezondheidswinkel,* Ankh Hermes BV, Deventer, The Netherlands.

Celia Wookey — *Myalgic Encephalomyelitis*, Croom Helm, London, Sydney, Dover, New Hampshire.
Konrad Lorenz — *Civilised Man's Eight Deadly Sins*, Methuen and Co. Ltd., London.
P. L. van der Harst — *Poging to Inleiding in die Practische Homoeopathie voor Artsen*, V.S.M. Geneesmiddelen BV, Alkmaar, The Netherlands.
Maurice Hanssen — *E for Additives*, Guild Publishing, London.
Doris Grant and Jean Joyce — *Food Combining for Health*, Thorsons Publishing Group, Wellingborough, England.